TEST BANK
FOR

Chemical Principles:
The Quest for Insight

THIRD EDITION

ROBERT J. BALAHURA

W. H. FREEMAN AND COMPANY
NEW YORK

ISBN: 0-7167-0739-X
EAN: 9780716707394

Printed in the United States of America

First printing

W. H. Freeman and Company
41 Madison Avenue
New York, NY 10010
Houndmills, Basingstoke RG21 6XS England

www.whfreeman.com

CONTENTS

Chapter 1: Atoms: The Quantum World

1. Visible light, microwaves, and x-rays travel through empty space at 3.00×10^8 m·s⁻¹.
   Ans: True

2. Which of the following statements is true regarding electromagnetic radiation?
   A) Electromagnetic radiation with a wavelength of 400 nm travels faster than that with a wavelength of 600 nm.
   B) The frequency of electromagnetic radiation determines how fast it travels.
   C) Electromagnetic radiation with a wavelength of 400 nm has a frequency that is smaller than that with a wavelength of 600 nm.
   D) Electromagnetic radiation with a wavelength of 600 nm travels faster than that with a wavelength of 400 nm.
   E) Electromagnetic radiation with a wavelength of 600 nm has a frequency that is smaller than that with a wavelength of 400 nm.
   Ans: E

3. What is the frequency of yellow light with a wavelength of 580 nm?
   A) 5.2×10^{16} Hz   D) 5.2×10^{12} Hz
   B) 1.9×10^{-17} Hz   E) 1.9×10^{-15} Hz
   C) 5.2×10^{14} Hz
   Ans: C

4. What is the wavelength of a radio station transmitting at 99.1 MHz?
   A) 330 nm   B) 303 nm   C) 0.00303 m   D) 3.03 m   E) 0.330 m
   Ans: D

5. Estimate the wavelength of the wave below.

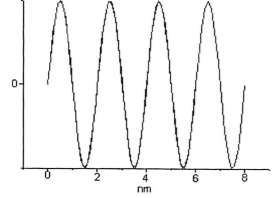

   A) 3 nm   B) 0.5 nm   C) 1 nm   D) 1.5 nm   E) 2 nm
   Ans: E

6. Arrange the following types of photons of electromagnetic radiation in order of decreasing energy: red light, radio, x-rays, γ-rays, infrared.
   A) γ-rays > x-rays > red light > infrared > radio
   B) radio > red light > infrared > x-rays > γ-rays
   C) x-rays > red light > infrared > γ-rays > radio
   D) x-rays > γ-rays > red light > infrared > radio
   E) γ-rays > x-rays > radio > infrared > red light
   Ans: A

7. In 1.0 s, a 60 W bulb emits 11 J of energy in the form of infrared radiation (heat) of wavelength 1850 nm. How many photons of infrared radiation does the lamp generate in 1.0 s?
   A) 1.0×10^{29} B) 1.0×10^{20} C) 6.8×10^{-14} D) 1.1×10^{-19} E) 6.6×10^{23}
   Ans: B

8. A lawyer who received a speeding ticket argues that because of the Heisenberg Uncertainty Principle the radar reading is uncertain. The judge, who happens to have a science degree, rules against the lawyer. Which of the following statements is true?
   A) The judge is incorrect because the uncertainty in position is large.
   B) The judge is correct because the car is so massive that the uncertainty in speed is very small.
   C) The judge is correct because the uncertainty in momentum is very large.
   D) The judge is incorrect because radar has only wave characteristics.
   E) The judge is incorrect because $(m\Delta v)(\Delta x) \geq \frac{1}{2}h$.
   Ans: B

9. Calculate the wavelength of a motorcycle of mass 275 kg traveling at a speed of 125 km·hr^{-1}.
   A) 6.94×10^{-38} m   D) 2.41×10^{-36} m
   B) 1.93×10^{-38} m   E) 2.08×10^{-29} m
   C) 1.93×10^{-41} m
   Ans: A

10. You are caught in a radar trap and hope to show that the speed measured by the radar gun is in error due to the uncertainty principle. If you assume that the uncertainty in your position is large, say about 10 m, and that the car has a mass of 2150 kg, what is the uncertainty in the velocity?
    A) 1×10^{-33} m·s^{-1} D) 1×10^{38} m·s^{-1}
    B) 1×10^{-19} m·s^{-1} E) 1×10^{-38} m·s^{-1}
    C) 1×10^{33} m·s^{-1}
    Ans: E

11. In everyday life, we have no problem in measuring both the velocity and position of objects.
    Ans: True

12. The ionization energy of a sodium atom in the gas phase is 494 kJ·mol^{-1}. What is the maximum wavelength of light that will photoionize a sodium atom in the gas phase?
    A) 9.90×10^{-20} m   B) 1.21×10^{-9} m   C) 602 nm   D) 242 nm   E) 121 nm
    Ans: D

13. Calculate the velocity of an oxygen molecule if it has a de Broglie wavelength of 0.0140 nm.
    A)   890. m·s^{-1} D)   445 m·s^{-1}
    B) 3.00×10^8 m·s^{-1} E)   8.90 m·s^{-1}
    C)   1780 m·s^{-1}
    Ans: A

14. Which of the following experiments most directly supports de Broglie's hypothesis of the wave nature of matter?
    A)   blackbody radiation
    B)   the photoelectric effect
    C)   alpha-particle scattering by a metal foil
    D)   electron diffraction by a crystal
    E)   the emission spectrum of the hydrogen atom
    Ans: D

15. The Bohr radius for an electron in the ground state of a hydrogen atom is 52.9 pm. The Bohr radius for an electron in the n = 2 state of He$^+$ is
    A) 211.6 pm   B) 105.8 pm   C) 26.5 pm   D) 52.9 pm   E) 13.2 pm
    Ans: B

16. The first line (lowest energy) in the Balmer series appears at 15,233 cm^{-1}. The second line appears at
    A)   109,678 cm^{-1} D)   30,466 cm^{-1}
    B)   45,699 cm^{-1} E)   20,311 cm^{-1}
    C)   20,565 cm^{-1}
    Ans: C

17. In a one-dimensional particle in a box, for n = 6, how many wavelengths equals the size of the box?
    A) 0   B) 3   C) 1   D) 12   E) 6
    Ans: B

18. In a one-dimensional particle in a box, for Ψ_4, how many nodes are predicted?
    A) 1   B) 3   C) 0   D) 2   E) 4
    Ans: B

19. In a one-dimensional particle in a box, the *zero-point energy* corresponds to
    A)   a node.
    B)   n = 0.
    C)   n = 1.
    D)   a quantum state where the Uncertainty Principle is not valid.
    E) $\Psi^2 = 0$.
    Ans: C

20. If an electron is confined to a one-dimensional box 200 pm in length, calculate the *zero-point energy*.
    A) 1.51×10^{-18} J   D) 1.21×10^{-19} J
    B) 6.04×10^{-18} J   E) 1.21×10^{-17} J
    C) 1.51×10^{-20} J
    Ans: A

21. If an electron is confined to a one-dimensional box 200 pm in length, calculate the wavelength of light required to promote the electron from the ground state to the first excited state.
    A) 329 pm   B) 263 pm   C) 165 pm   D) 439 pm   E) 1320 pm
    Ans: D

22. Which one of the following statements is incorrect?
    A)   For a one-dimensional particle in a box, as the mass of the particle becomes larger the separation between neighboring energy levels increases.
    B)   For a one-dimensional particle in a box, the separation between neighboring energy levels decreases as the length of the container increases.
    C)   For a one-dimensional particle in a box, the separation between neighboring energy levels becomes zero when the walls of the box are infinitely far apart.
    D)   Argon atoms in a cylinder can be treated as though their translational energy were not quantized.
    E)   A billiard ball on a table has a completely negligible *zero-point energy*.
    Ans: A

23. If a particle is confined to a one-dimensional box of length 300 pm, for Ψ_3 the particle has zero probability of being found at
    A)   100 and 200 pm, respectively.
    B)   150 pm only.
    C)   50, 150, and 250 pm, respectively.
    D)   50 and 250 pm, respectively.
    E)   75, 125, 175, and 225 pm, respectively.
    Ans: A

24. If a particle is confined to a one-dimensional box of length 300 pm, for Ψ_3 the particle is most likely to be found at
    A)   0 pm.
    D)   300 pm.
    B)   50, 150, and 250 pm, respectively.
    E)   100 and 200 pm, respectively.
    C)   17.3 pm.
    Ans: B

25. What is the probability of finding an electron in a small region of a hydrogen 1s orbital at a distance $3a_0$ from the nucleus relative to the probability of finding it in the same small region located at the nucleus?
    A) 14%   B) 0.25%   C) 22%   D) 1.8%   E) 5.0%
    Ans: B

26. Which of the following emission lines corresponds to part of the Balmer series of lines in the spectrum of a hydrogen atom?
      a. $n_2 \rightarrow n_1$
      b. $n_4 \rightarrow n_2$
      c. $n_4 \rightarrow n_1$
      d. $n_3 \rightarrow n_2$
      e. $n_4 \rightarrow n_3$
    A) b and d   B) a, d, and e   C) a and c   D) e   E) b and c
    Ans: A

27. Calculate the longest-wavelength line in the Balmer series for hydrogen.
    A) 182 nm   B) 657 nm   C) 536 nm   D) 122 nm   E) 486 nm
    Ans: B

28. The Balmer series of lines for the hydrogen atom are found only in the visible region of the spectrum.
    Ans: False

29. What is the shortest-wavelength line in the emission spectrum of the hydrogen atom?
    A) 182 nm   B) 100 nm   C) 122 nm   D) 91.2 nm   E) 1.00 nm
    Ans: D

30. The total number of orbitals in a shell with principal quantum number 5 is
    A) 32.   B) 50.   C) 25.   D) 40.   E) 5.
    Ans: C

31. Which set of quantum numbers, n, l, m_l, could correspond to one of the highest energy electrons in Zr?
    A) 4, 2, –2   B) 4, 2, +3   C) 3, 2, –2   D) 4, 3, –2
    Ans: A

32. Which set of quantum numbers could correspond to a 4f orbital?
    A) $n = 4, l = 4, m_l = +3$ D) $n = 3, l = 2, m_l = +1$
    B) $n = 4, l = 3, m_l = +4$ E) $n = 3, l = 2, m_l = 0$
    C) $n = 4, l = 3, m_l = -3$
    Ans: C

33. How many total nodal planes are present in the 3d orbitals?
    A) 15   B) 0   C) 5   D) 20   E) 10
    Ans: E

34. How many nodes are present in a 3s and a 3p orbital, respectively?
    A) 2 and 1   B) 0 and 1   C) 0 and 2   D) 1 and 1   E) 2 and 2
    Ans: E

35. Nodal planes in orbitals can be accounted for by the wavelike behavior of electrons.
    Ans: True

36. How many *nodal planes* are present in an f orbital?
    A) 2   B) 3   C) 7   D) 4   E) 5
    Ans: B

37. Which of the following is true?
    A)   A 2s orbital has one nodal plane.
    B)   An electron in a p-orbital has zero probability of being found at the nucleus.
    C)   A p-orbital has a spherical boundary surface.
    D)   An s-orbital becomes more dense as the distance from the nucleus increases.
    E)   An electron in an s-orbital has a zero probability of being found at the nucleus.
    Ans: B

38. The three quantum numbers for an electron in a hydrogen atom in a certain state are
    $n = 4, l = 1, m_l = 1$. The electron is located in what type of orbital?
    A) 4s   B) 3p   C) 3d   D) 4d   E) 4p
    Ans: E

39. The three quantum numbers for an electron in a hydrogen atom in a certain state are
    $n = 4, l = 2, m_l = 1$. The electron is located in what type of orbital?
    A) 4p   B) 3p   C) 4s   D) 4d   E) 3d
    Ans: D

40. Which of the following is true?
    A) A p-electron penetrates more than an s-electron through the inner shells of an atom.
    B) A p-electron penetrates less than a d-electron through the inner shells of an atom.
    C) A p-electron has a non-zero probability density for finding it at the nucleus.
    D) A d-electron has a non-zero probability density of finding it at the nucleus.
    E) A p-electron experiences a smaller effective nuclear charge than an s-electron.
    Ans: E

41. Write the ground-state electron configuration of a chromium atom.
    A) $[Ar]4s^2 3d^4$ B) $[Ar]4s^1 3d^5$ C) $[Ar]3d^5 4s^1$ D) $[Ar]3d^6$ E) $[Ar]3d^4 4s^2$
    Ans: C

42. Write the ground-state electron configuration of a europium atom.
    A) $[Xe]5d^7 6s^2$ D) $[Xe]4f^7 6s^2$
    B) $[Xe]4f^2 5d^5 6s^2$ E) $[Xe]4f^5 5d^2 6s^2$
    C) $[Xe]4f^9$
    Ans: D

43. Write the ground-state electron configuration of a terbium atom.
    A) $[Xe]4f^{11}$ D) $[Xe]4f^4 5d^5 6s^2$
    B) $[Xe]4f^{10} 6s^1$ E) $[Xe]4f^6 5d^5$
    C) $[Xe]4f^9 6s^2$
    Ans: C

44. Write the ground-state electron configuration of a lead atom.
    A) $[Xe]4f^{14} 5d^5 6s^1 6p^6 7s^2$ D) $[Xe]4f^{14} 5d^{10} 6p^4$
    B) $[Xe]4f^{14} 5d^{10} 6s^2 6p^2$ E) $[Xe]4f^{14} 5d^9 6s^2 6p^3$
    C) $[Xe]4f^{14} 5d^{10} 6s^1 6p^3$
    Ans: B

45. Write the ground-state electron configuration of a ruthenium atom.
    A) $[Kr]4d^6 5s^2$ B) $[Kr]4d^7 5p^1$ C) $[Kr]4d^5 5p^3$ D) $[Kr]4d^8$ E) $[Kr]4d^7 5s^1$
    Ans: E

46. Write the ground-state electron configuration of a samarium atom.
    A) $[Xe]4f^7 5d^1$ B) $[Xe]5d^8$ C) $[Xe]4f^7 6s^1$ D) $[Xe]4f^8$ E) $[Xe]4f^6 6s^2$
    Ans: E

47. Write the ground-state electron configuration of a tin(IV) ion.
    A) $[Kr]4d^3 5s^1 5p^6$ D) $[Kr]4d^5 5p^5$
    B) $[Kr]4d^4 5p^6$ E) $[Kr]4d^{10}$
    C) $[Kr]4d^5 5s^2 5p^3$
    Ans: E

48. Write the ground-state electron configuration of In^+.
   A) $[Kr]4d^75s^25p^3$
   B) $[Kr]4d^85s^15p^3$
   C) $[Kr]4d^55s^15p^6$
   D) $[Kr]4d^{10}5s^2$
   E) $[Kr]4d^{10}5s^15p^1$
   Ans: D

49. Write the ground-state electron configuration of Tl^+.
   A) $[Xe]4f^{14}5d^{10}6p^2$
   B) $[Xe]4f^{14}5d^{10}6s^2$
   C) $[Xe]4f^{14}5d^{10}6s^16p^1$
   D) $[Xe]4f^{14}5d^86s^16p^3$
   E) $[Xe]4f^{14}5d^56s^16p^6$
   Ans: B

50. Write the ground-state electron configuration of Pb^{2+}.
   A) $[Xe]4f^{14}5d^56s^16p^6$
   B) $[Xe]4f^{14}5d^{10}6s^2$
   C) $[Xe]4f^{14}5d^56s^26p^5$
   D) $[Xe]4f^{14}5d^{10}6s^16p^1$
   E) $[Xe]4f^{14}5d^{10}6p^2$
   Ans: B

51. Write the ground-state electron configuration of Sb^{3+}.
   A) $[Kr]4d^85s^15p^3$
   B) $[Kr]4d^55s^15p^6$
   C) $[Kr]4d^{10}5p^2$
   D) $[Kr]4d^{10}5s^15p^1$
   E) $[Kr]4d^{10}5s^2$
   Ans: E

52. All of the following can have the ground-state electron configuration $[Xe]4f^{14}5d^{10}$ except
   A) Pb^{4+} B) Hg^{2+} C) Bi^{5+} D) Tl^+ E) Au^+
   Ans: D

53. All of the following can have the ground-state electron configuration $[Kr]4d^{10}$ except
   A) Cd^{2+} B) Ag^+ C) Pd D) In^+ E) Sn^{4+}
   Ans: D

54. Which of the following has the smallest atomic radius?
   A) Cl B) P C) S D) Si E) Al
   Ans: A

55. Which of the following has the largest radius?
   A) S^{2-} B) Cl C) Cl^- D) K^+ E) S
   Ans: A

56. Which of the following species is isoelectronic with S^{2-}?
   A) Mg^{2+} B) Rb^+ C) Ar D) As^{3-} E) Br^-
   Ans: C

57. Which of the following species is isoelectronic with Kr?
A) K^+ B) Cl^- C) Ar   D) Xe   E) Sr^{2+}
Ans: E

58. From the data below, which element is likely to be a metal?

Element	First Ionization Energy, $kJ \cdot mol^{-1}$
1	1310
2	1011
3	418
4	2080
5	947

A) 2   B) 5   C) 3   D) 1   E) 4
Ans: C

59. From the data below, which elements are likely to be nonmetals?

Element	First Ionization Energy, $kJ \cdot mol^{-1}$
1	1310
2	980
3	418
4	2080
5	947

A) 3 and 5   B) 3 only   C) 1 and 4   D) 1 and 2   E) 2 and 5
Ans: C

60. Which of the following is likely to form ions two units lower in charge than expected from the group number?
A) Tl   B) Hg   C) Zn   D) Se   E) Cd
Ans: A

61. Which of the following is likely to form ions two units lower in charge than expected from the group number?
A) Hg   B) Cd   C) Sb   D) Ge   E) Zn
Ans: C

62. Given the following elements and three values of possible first ionization energies:
    Cl, Ge, K and 418, 1255, 784 kJ·mol^{-1}.
    Match the atoms with their first ionization energies.
    A)   Cl (418), Ge (784), and K (1255 kJ·mol^{-1})
    B)   Cl (1255), Ge (784), and K (418 kJ·mol^{-1})
    C)   Cl (784), Ge (1255), and K (418 kJ·mol^{-1})
    D)   Cl (1255), Ge (418), and K (784 kJ·mol^{-1})
    E)   Cl (418), Ge (1255), and K (784 kJ·mol^{-1})
    Ans: B

63. Write an equation that represents the second ionization energy of copper.
    Ans: $Cu^+(g) \rightarrow Cu^{2+}(g) + e^-(g)$

64. Consider the following ground-state electronic configurations. Which atom has both the highest first ionization energy and the highest electron affinity?
    A) [Ne] $3s^23p^5$ B) [Ne] $3s^23p^3$ C) [Ne] $3s^23p^1$ D) [Ne] $3s^23p^4$
    Ans: A

65. In each pair, which ionization reaction is largest?
    (a) I_3 of B or I_3 of Be
    (b) I_4 of C or I_3 of B
    Ans: (a) I_3 of Be
          (b) I_4 of C

66. What is the subshell notation and the number of orbitals having the quantum numbers $n = 4$, $l = 3$?
    A) 4d and 5   B) 4p and 3   C) 3f and 7   D) 3d and 5   E) 4f and 7
    Ans: E

67. What is the subshell notation and the number of orbitals having the quantum numbers $n = 4$, $l = 2$?
    A) 4d and 10   B) 4f and 14   C) 4d and 5   D) 4p and 3   E) 4f and 7
    Ans: C

68. Which of the following would be most reactive with air and water?
    A) Ba   B) Mg   C) Ga   D) Br
    Ans: A

69. Which of the following has similar properties to Al?
    A) Li   B) Be   C) Si   D) Ga   E) Mg
    Ans: B

70. Which of the following elements have the most metal character?
    A) In   B) Ge   C) Te   D) I
    Ans: A

71. All of the following are metalloids except
    A) B   B) As   C) Ge   D) Sb   E) Si
    Ans: A

72. How many *nodal planes* are expected for 2s and 4f orbitals, respectively?
    A) 0 and 4   B) 0 and 3   C) 2 and 4   D) 1 and 4   E) 1 and 3
    Ans: B

73. Which of the following subshells cannot exist in an atom?
    A) 4d   B) 5g   C) 5f   D) 4f   E) 3f
    Ans: E

74. What is the **inert-pair effect**?
    Ans: The inert-pair effect is the tendency to form ions two units lower in charge than expected from the group number.

75. When an electron is added to a gaseous chlorine atom, 349 kJ·mol^{-1} of energy are released. What is the ionization energy of a gaseous chloride ion?
    Ans: +349 kJ·mol^{-1}

76. Which of the following is correct with respect to the photoelectric effect?
    A)   A plot of the kinetic energy of the ejected electrons versus the frequency of the incident radiation has a slope that is equal to the value of the work function.
    B)   All metals have the same work function.
    C)   The kinetic energy of the ejected electrons increases with the intensity of the incident radiation.
    D)   A plot of the kinetic energy of the ejected electrons versus the frequency of the incident radiation is linear.
    Ans: D

77. Which of the following is expected to approximate the value of the wavelength of a 5-g bullet traveling at 1300 miles per hour?
    A) 2 m   B) 2×10^{-34} m   C) 200 pm   D) 2 pm
    Ans: B

78. Estimate the minimum uncertainty in the position of an electron of mass 9.109×10^{-31} kg given that its speed is unknown to within $\pm 3.00 \times 10^5$ m.s^{-1}.
    A) 386 pm   B) 386×10^{-12} m   C) 193 pm   D) 1.93×10^{-12}
    Ans: C

79. The four π electrons of 1,3-butadiene can be modeled as particles in a box. If the length of 1,3-butadiene (L) is 0.56 nm, calculate the energy of the lowest energy transition.
    Ans: 9.6×10^{-19} J

80. Carotene, which is partly responsible for the color of carrots, has 22 π electrons in a conjugated system and a length of about 3 nm. The lowest energy transition in carotene is from $n =$ ___ to $n =$ ___.
    Ans: 11, 12

81. Comparing the lowest energy transition, ΔE, in 1,3-butadiene(A) and 1,3,5,7-octatetraene (B), ΔE_A is greater than ΔE_B. True or False?
    Ans: True

82. The zero point energy in 1,3-butadiene is larger than in 1,3,5,7-octatetraene. True or False?
    Ans: True

83. Which of the following atoms has the highest electron affinity?
    A) Ar   B) P   C) Al   D) Si
    Ans: D

84. Because of fluorine's high electronegativity, it requires less energy to make F^{2-} from F^- than to make O^{2-} from O^-. True of False?
    Ans: False

85. Which of the following metals has the lowest work function?
    A) Be   B) Sr   C) Li   D) Rb
    Ans: D

Chapter 2: Chemical Bonds

1. Use the expression for the Coulomb potential energy to calculate the energy for formation of 1 mole of sodium chloride ion-pairs, i.e., the energy change for the following reaction:

    $$Na^+(g) + Cl^-(g) \rightarrow Na^+Cl^-(g)$$

    Use $r_{12} = 283$ pm.
    Ans: -491 kJ·mol^{-1}

2. If 491 kJ·mol^{-1} are released in the reaction $Na^+(g) + Cl^-(g) \rightarrow Na^+Cl^-(g)$, what is the energy change for the reaction $Na(g) + Cl(g) \rightarrow Na^+Cl^-(g)$? (Hint: See the discussion in the text and apply Hess's Law.)
    Ans: -346 kJ·mol^{-1}

3. If 346 kJ·mol^{-1} are released in the reaction $Na(g) + Cl(g) \rightarrow Na^+Cl^-(g)$, the energy change for the reaction $Na^+Cl^-(g) \rightarrow NaCl(s)$ is *endothermic or exothermic*?
    Ans: exothermic

4. The Madelung constant is different for all crystals. True or False?
    Ans: True

5. Use the expression for the Coulomb potential energy to calculate the energy for formation of 1 mole of rubidium chloride ion-pairs, i.e., the energy change for the following reaction:

    $$Rb^+(g) + Cl^-(g) \rightarrow Rb^+Cl^-(g)$$

    Use $r_{12} = 330$ pm.
    Ans: -421 kJ·mol^{-1}

6. Which of the following has the lowest lattice energy?
    A) KCl   B) LiCl   C) KBr   D) NaCl   E) KI
    Ans: E

7. Which of the following has the highest lattice energy?
    A) NaCl   B) KI   C) MgO   D) BaO   E) CaO
    Ans: C

8. Which of the following would have the highest melting point?
    A) KF   B) KI   C) RbF   D) KBr   E) KCl
    Ans: A

9. Nonmetals rarely lose electrons in chemical reactions because
   A) their electron affinities are too high.
   B) their ionic radii become too small.
   C) their ionization energies are too small.
   D) their size is to small.
   E) their ionization energies are too high.
   Ans: E

10. An element, E, has the electronic configuration [Ne] $3s^2 3p^1$. Write the formula of its compound with sulfate.
    Ans: $E_2(SO_4)_3$

11. Predict the electronic configuration in the phosphide ion in Ca_3P_2.
    A) $[Ne]3s^2 3p^6$ or [Ar]   D) $[Ne]3s^1 3p^3$
    B) $[Ne]3s^2 3p^5$ E) $[Ne\}3s^2 3p^3$
    C) $[Ne]3s^2 3p^6 4s^2$
    Ans: A

12. Write the formula of calcium nitride.
    Ans: Ca_3N_2

13. Which of the following metal ions has the ground-state electron configuration $[Ar]3d^6$?
    A) Ni^{3+} B) Fe^{2+} C) Mn^{2+} D) Cu^+ E) Ca^{2+}
    Ans: B

14. For the ground-state ion Pb^{2+}, what type of orbital do the electrons with highest energy reside in?
    A) 6p   B) 5p   C) 4f   D) 6s   E) 5d
    Ans: D

15. For the ground-state ion Pb^{4+}, what type of orbital do the electrons with highest energy reside in?
    A) 5p   B) 6p   C) 4f   D) 5d   E) 6s
    Ans: D

16. For the ground-state ion Bi^{3+}, what type of orbital do the electrons with highest energy reside in?
    A) 5d   B) 6s   C) 4f   D) 5p   E) 6p
    Ans: B

17. For the ground-state ion I^-, what type of orbital do the electrons with highest energy reside in?
    A) 4d   B) 6s   C) 5p   D) 5d   E) 5s
    Ans: C

18. Because of the octet rule, the gaseous O^{2-} ion is stable.
    Ans: False

19. All of the following elements exist as diatomic gases at room temperature and atmospheric pressure except
    A) H   B) Ar   C) N   D) Cl   E) O
    Ans: B

20. How many lone pairs of electrons are found in the Lewis structure of the interhalogen compound BrCl?
    A) 6   B) 2   C) 7   D) 3   E) 4
    Ans: A

21. How many lone pairs of electrons are found in the Lewis structure of urea, $(NH_2)_2CO$?
    A) 2   B) 3   C) 6   D) 4   E) 8
    Ans: D

22. How many lone pairs of electrons are found in the Lewis structure of hydrazine, H_2NNH_2?
    A) 8   B) 4   C) 1   D) 0   E) 2
    Ans: E

23. Draw the Lewis structure of xenon difluoride and give the number of lone pairs electrons around the central atom.
    Ans: three lone pairs

24. Draw the Lewis structure of the formate ion and indicate if resonance forms are possible.
    Ans: Two resonance forms are possible.

25. Draw the "best" Lewis structures of hydrogen azide, $HN_1N_2N_3$, and the azide ion, $N_1N_2N_3-$. The subscripts are used for identification. Match the following bond lengths to the correct N–N bond. The bond lengths can be used more than once.

	N–N bond	bond length, pm
hydrogen azide	N_1–N_2	113
	N_2–N_3	116
azide ion	N_1–N_2	124
	N_2–N_3	

    Ans: hydrogen azide: N_1–N_2, 124 pm, N_2–N_3, 113 pm; azide ion: N_1–N_2, 116 pm, N_2–N_3, 116 pm

26. All of the following have resonance structures except
    A) H_2CO B) CH_3CONH^- C) $CH_2COCH_3^-$ D) $CH_2CHCH_2^+$
    Ans: A

27. For dinitrogen monoxide, the arrangement of the atoms is N-N-O. In the Lewis structure with a double bond between NN and NO, the formal charges on N, N, and O, respectively, are
    A) 0, −1, +1   B) −1, +1, 0   C) 0, +1, −1   D) 0, 0, 0   E) −2, +1, +1
    Ans: B

28. For dinitrogen monoxide, the arrangement of the atoms is N-N-O. In the Lewis structure with a single bond between NN and a triple bond between NO, the formal charges on N, N, and O, respectively, are
    A) −1, +1, 0   B) 0, 0, 0   C) 0, +1, −1   D) 0, −1, +1   E) −2, +1, +1
    Ans: E

29. In the "best" Lewis structure of XeO_4, there are two double bonds and the formal charge on Xe is zero. True or False?
    Ans: False

30. Write three Lewis structures for the cyanate ion, NCO^-, where the arrangement of atoms is N-C-O. In the most plausible structure,
    A)   there is a triple bond between N and C.
    B)   there are two double bonds.
    C)   there is a triple bond between C and O.
    D)   the formal charge on O is +1.
    E)   the formal charge on N is −1.
    Ans: A

31. Predict the N-O bond lengths in NO_2^-, given the N-O and N=O bond lengths of 140 and 120 pm, respectively.
    Ans: Both ~ 130 pm

32. Why are the N-O bond lengths in NO_3^- the same?
    Ans: This can be explained by resonance.

33. Which of the following species are radicals?
    A) NO_2 B) HNO_2 C) N_2O D) CH_2O E) HCN
    Ans: A

34. Which of the following species are radicals?
    A) CH_2O B) HCN   C) HClO   D) $ClONO_2$ E) ClO
    Ans: E

35. In the most plausible Lewis structure of $XeOF_2$, there are
    A)   2 single bonds, 1 double bond, and 1 lone pair of electrons around Xe.
    B)   3 single bonds and 1 lone pair of electrons around Xe.
    C)   2 single bonds, 1 double bond, and 3 lone pairs of electrons around Xe.
    D)   2 single bonds, 1 double bond, and 2 lone pairs of electrons around Xe.
    E)   3 single bonds and 2 lone pairs of electrons around Xe.
    Ans: D

36. How many electrons are in the expanded valence in $XeOF_2$?
    A) 14   B) 12   C) 8   D) 10   E) 6
    Ans: B

37. How many electrons are in the expanded valence in I_3^-?
    A) 12   B) 6   C) 10   D) 14   E) 8
    Ans: C

38. How many electrons are in the expanded valence in SO_2?
    A) 10   B) 14   C) 8   D) 6   E) 12
    Ans: A

39. How many electrons are in the expanded valence in XeO_4?
    Ans: 16

40. Consider the following equilibrium:
    $$S_2O_4^{2-}(aq) \rightleftharpoons 2SO_2^-(aq) \qquad K \sim 10^{-9}$$
    Write a Lewis structure for each species.
    Ans: The arrangement of atoms in $S_2O_4^{2-}$ is O_2S—SO_2. The latter has a Lewis
          structure that obeys the octet rule, but SO_2^- is a radical.

41. Which of the following species has bonds with the most ionic character?
    A) SiO_2 B) PCl_3 C) P_4O_{10} D) CO_2 E) NO_2
    Ans: A

42. Write all possible Lewis structures of sulfur dioxide. Which structure is most feasible?
    Ans: The structure with the expanded valence is favored.

43. Which of the following species has bonds with the most ionic character?
    A) CO_2 B) NO_2 C) SnO_2 D) P_4O_{10} E) PCl_3
    Ans: C

44. Which of the following statements is true?
    A) Atoms with high ionization energies and high electron affinities are highly electronegative.
    B) Atoms with high ionization energies and high electron affinities have low electronegativities.
    C) The electronegativity of an atom depends only on the value of the ionization energy of the atom.
    D) Atoms with low ionization energies and low electron affinities have high electronegativities.
    E) The electronegativity of an atom is defined as ½(Electron Affinity) of the atom.
    Ans: A

45. Which of the following statements is true?
    A) The electronegativity of an atom is defined as ½(Electron Affinity) of the atom.
    B) The electronegativity of an atom depends only on the value of the ionization energy of the atom.
    C) Atoms with high ionization energies and high electron affinities have low electronegativities.
    D) Atoms with low ionization energies and low electron affinities have low electronegativities.
    E) Atoms with low ionization energies and low electron affinities have high electronegativities.
    Ans: D

46. Which of the compounds below has bonds with the least covalent character?
    A) AgI   B) AgCl   C) AgF   D) AlCl$_3$ E) BeCl$_2$
    Ans: C

47. Which of the compounds below has bonds with the most covalent character?
    A) NaCl   B) LiCl   C) CaCl$_2$ D) BeCl$_2$ E) MgCl$_2$
    Ans: D

48. Which of the compounds below has bonds with the most covalent character?
    A) CaO   B) Li$_2$O   C) MgO   D) MgS   E) CaS
    Ans: D

49. Use the bond enthalpies given to estimate the heat released when 1-bromobutene, $CH_3CH_2CH=CH_2$, reacts with bromine to give $CH_3CH_2CHBrCH_2Br$. Bond enthalpies (kJ·mol^{-1}): C-H, 412; C-C, 348; C=C, 612; C-Br, 276; Br-Br, 193.
    A) 181 kJ·mol^{-1} D) 95 kJ·mol^{-1}
    B) 317 kJ·mol^{-1} E) 507 kJ·mol^{-1}
    C) 288 kJ·mol^{-1}
    Ans: D

50. Use the bond enthalpies given to estimate the heat released when ethene, $CH_2=CH_2$, reacts with HBr to give CH_3CH_2Br. Bond enthalpies (kJ·mol^{-1}): C-H, 412; C-C, 348; C=C, 612; C-Br, 276; Br-Br, 193; H-Br, 366.
    A)   1036 kJ·mol^{-1} D)   424 kJ·mol^{-1}
    B)   200 kJ·mol^{-1} E)   58 kJ·mol^{-1}
    C)   470 kJ·mol^{-1}
    Ans: E

51. Use the bond enthalpies given to estimate the heat released when 2-methyl-1-propene, $(CH_3)_2C=CH_2$, reacts with HBr to give $(CH_3)_2CBrCH_3$. Bond enthalpies (kJ·mol^{-1}): C-H, 412; C-C, 348; C=C, 612; C-Br, 276; H-Br, 366.
    A)   58 kJ·mol^{-1} D)   288 kJ·mol^{-1}
    B)   507 kJ·mol^{-1} E)   181 kJ·mol^{-1}
    C)   317 kJ·mol^{-1}
    Ans: A

52. Use the bond enthalpies given to estimate the heat released when ethene, $CH_2=CH_2$, reacts with hydrogen to give CH_3CH_3. Bond enthalpies (kJ·mol^{-1}): C-H, 412; C-C, 348; C=C, 612; C-Br, 276; H-H, 436.
    A)   124 kJ·mol^{-1} D)   148 kJ·mol^{-1}
    B)   342 kJ·mol^{-1} E)   560 kJ·mol^{-1}
    C)   288 kJ·mol^{-1}
    Ans: A

53. Which of the following compounds contains the weakest bonds to hydrogen?
    A) CH_4 B) H_2O C) SiH_4 D) HF   E) H_2S
    Ans: C

54. Which of the following compounds contains the strongest bonds to hydrogen?
    A) SiH_4 B) CH_4 C) HF   D) H_2S E) H_2O
    Ans: C

55. Which of the following compounds is the least stable?
    A) CH_4 B) SnH_4 C) SiH_4 D) GeH_4 E) PbH_4
    Ans: E

56. Estimate the CO bond length in acetone, CH_3COCH_3. Given: covalent radii (pm) of C–, 77; C=, 67; O–, 74; O=, 60; H, 37.
    A) 75.5 pm   B) 127 pm   C) 63.5 pm   D) 151 pm   E) 137 pm
    Ans: B

57. Estimate the CN bond length in urea, NH_2CONH_2. Given: covalent radii (pm) of C–, 77; C=, 67; N–, 75; N=, 60; O–, 74; O=, 60; H, 37.
    A) 71 pm   B) 127 pm   C) 76 pm   D) 152 pm   E) 142 pm
    Ans: D

58. If the following crystallize in the same type of structure, which has the highest lattice energy?
    A) LiCl   B) KF   C) KBr   D) KCl   E) LiF
    Ans: E

59. If the following crystallize in the same type of structure, which has the highest lattice energy?
    A) NaCl   B) NaF   C) KF   D) NaBr   E) NaI
    Ans: B

60. If the following crystallize in the same type of structure, which has the lowest lattice energy?
    A) CaO   B) BaS   C) SrO   D) SrS   E) BaO
    Ans: B

61. If the following crystallize in the same type of structure, which has the lowest lattice energy?
    A) LiCl   B) NaI   C) NaCl   D) KCl   E) KI
    Ans: E

62. White phosphorus is composed of tetrahedral molecules of P_4 in which every P atom is connected to three other P atoms. In the Lewis structure of P_4, there are
    A)   3 bonding pairs and 4 lone pairs of electrons.
    B)   6 bonding pairs and 2 lone pairs of electrons.
    C)   5 bonding pairs and 4 lone pairs of electrons.
    D)   6 bonding pairs and no lone pairs of electrons.
    E)   6 bonding pairs and 4 lone pairs of electrons.
    Ans: E

63. Which of the following is a radical?
    A) BrO   B) CH_3^+ C) CH_3^- D) BF_4^-
    Ans: A

64. If dinitrogen oxide has a dipole moment, what is the arrangement of atoms?
    Ans: N-N-O

65. The electronegativity of an element can be expressed as $\frac{1}{2}(I + E_a)$ where I is the ionization energy and E_a is the electron affinity.
    Ans: True

66. The best Lewis structures of SO_2 and O_3 include expanded valence structures such as O=S=O and O=O=O.
    Ans: False

67. Which of the following has resonance structures?
    A) $XeOF_2$ B) N_2H_4 C) CH_3CONH^- D) H_2CO
    Ans: C

68. How many resonance structures can be drawn for N_2O?
    A) 0   B) 3   C) 2   D) 1
    Ans: B

69. What is the formal charge on the Xe atom in XeF_4?
    A) 0   B) −4   C) +2   D) +4
    Ans: A

70. There are three resonance structures of the sulfate ion. A resonance structure can be written where the formal charge on sulfur is 0. True or False?
    Ans: True

71. How many double bonds are present in the "best" resonance structure of the phosphate ion?
    A) 2   B) 3   C) 1   D) 0
    Ans: C

72. How many lone pairs of electrons are there in the Lewis structure of Al_2Cl_6?
    A) 24   B) 12   C) 4   D) 16
    Ans: D

73. Match the following compounds with their lattice energy.
    KI, LiF, MgF_2, LiI   2961, 1046, 759, 645 kJ/mol
    Ans: MgF_2 (2961), LiF (1046), LiI (759), KI (645 kJ/mol)

74. White phosphorus is composed of tetrahedral molecules of P_4 in which each P atom is bonded to three others. In this molecule the formal charge on each P atom is ___.
    Ans: 0

75. In the following molecules which bonds are the strongest?
    A) H_2O B) H_2Se C) H_2Te D) H_2S
    Ans: A

76. An element E has the electronic configuration $1s^2 2s^2 2p^4$. What is the formula of its compound with lithium?
    A) LiE_2 B) LiE   C) Li_2E D) Li_4E
    Ans: C

Chapter 3: Molecular Shape and Structure

1. What are all the angles in a trigonal bipyramidal geometry?
   Ans: 90°, 120°, and 180°

2. What are all the angles in a trigonal planar geometry?
   Ans: 120°

3. Predict the electron arrangement in NH_2^-.
   Ans: tetrahedral

4. Predict the electron arrangement in NO_2^-.
   Ans: trigonal planar

5. Predict the electron arrangement in IF_4^+.
   Ans: trigonal bipyramidal

6. Predict the electron arrangement in ClF_3.
   Ans: trigonal bipyramidal

7. Predict the electron arrangement in IF_5.
   Ans: octahedral

8. What is the shape of AlH_4^-?
   A)   tetrahedral   D)   T-shaped
   B)   trigonal bipyramidal   E)   square planar
   C)   seesaw
   Ans: A

9. What is the shape of BrO_4^-?
   A)   tetrahedral   D)   T-shaped
   B)   trigonal bipyramidal   E)   square planar
   C)   seesaw
   Ans: A

10. What is the shape of AsF_3?
    A)   T-shaped   D)   tetrahedral
    B)   trigonal planar   E)   seesaw
    C)   trigonal pyramidal
    Ans: C

11. What is the shape of SO_3^{2-}?
    A) T-shaped
    B) trigonal pyramidal
    C) seesaw
    D) tetrahedral
    E) trigonal planar
    Ans: B

12. What is the shape of CS_3^{2-}?
    A) trigonal pyramidal
    B) trigonal planar
    C) T-shaped
    D) tetrahedral
    E) seesaw
    Ans: B

13. What is the shape of $COCl_2$?
    A) T-shaped
    B) trigonal planar
    C) trigonal pyramidal
    D) tetrahedral
    E) seesaw
    Ans: B

14. What is the shape of XeF_4?
    A) square planar
    B) tetrahedral
    C) trigonal bipyramidal
    D) seesaw
    E) T-shaped
    Ans: A

15. What is the shape of ICl_4^-?
    A) T-shaped
    B) trigonal bipyramidal
    C) seesaw
    D) tetrahedral
    E) square planar
    Ans: E

16. What is the shape of IF_4^+?
    A) tetrahedral
    B) seesaw
    C) trigonal bipyramidal
    D) square planar
    E) T-shaped
    Ans: B

17. What is the shape of ClF_3?
    A) tetrahedral
    B) seesaw
    C) trigonal bipyramidal
    D) T-shaped
    E) square planar
    Ans: D

18. All of the following have a linear shape except
    A) CH_2^{2-}. B) O_3. C) I_3^-. D) COS. E) CS_2.
    Ans: B

19. All of the following have an angular shape except
    A) HOCl.   B) S_3^{2-}.   C) I_3^-.   D) ClO_2^-.   E) NH_2^-.
    Ans: C

20. All of the following have a linear shape except
    A) CH_2^{2-}.   B) CS_2.   C) COS.   D) I_3^-.   E) I_3^+.
    Ans: E

21. All of the following have an angular shape except
    A) N_3^-.   B) ClO_2^-.   C) S_3^{2-}.   D) HOCl.   E) NH_2^-.
    Ans: A

22. Which of the following has bond angles slightly less than 109°?
    A) NH_4^+ B) ClO_4^- C) BrO_3^- D) PO_4^{3-} E) BH_4^-
    Ans: C

23. Which of the following has bond angles of 180°?
    A) I_3^- B) ClO_2^- C) O_3 D) NH_2^- E) HO_2^-
    Ans: A

24. Which of the following has bond angles of 180°?
    A) N_2O B) ClO_2^- C) O_3 D) HO_2^- E) NH_2^-
    Ans: A

25. Which of the following has bond angles slightly less than 120°?
    A) SO_3 B) SF_2 C) I_3^- D) NO_3^- E) O_3
    Ans: E

26. Which of the following has bond angles slightly less than 109°?
    A) NO_2^- B) I_3^- C) HOCl   D) O_3 E) CH_2^-
    Ans: C

27. Which of the following has bond angles slightly less than 109°?
    A) CS_3^{2-} B) AsF_3 C) SO_2 D) $COCl_2$ E) COS
    Ans: B

28. Which of the following has bond angles slightly less than 120°?
    A) NO_3^- B) HO_2^- C) NO_2^- D) CS_3^{2-} E) I_3^+
    Ans: C

29. Which of the following has bond angles of 120°?
    A) HO_2^- B) CS_3^{2-} C) S_3^{2-} D) O_3 E) NO_2^-
    Ans: B

30. Which of the following has bond angles of 90°, 120°, and 180°?
    A) PF_6^- B) IF_5 C) XeF_4 D) ICl_4^- E) SF_4
    Ans: E

31. Which of the following only has bond angles of 90°?
    A) IF_5 B) IF_4^+ C) XeF_2 D) SF_4 E) IO_2F_3
    Ans: A

32. Which of the following only has bond angles of 90° and 180°?
    A) IF_5 B) BrF_3 C) BCl_3 D) NO_3^- E) ICl_4^+
    Ans: B

33. Which of the following is polar?
    A) SCN^- B) O_3 C) XeF_2 D) I_3^- E) N_2O
    Ans: B

34. Which of the following is polar?
    A) N_2O B) XeF_2 C) XeO_2 D) SCN^- E) I_3^-
    Ans: C

35. Which of the following is polar?
    A) XeF_4 B) PCl_5 C) ICl_4^- D) SF_6 E) IF_5
    Ans: E

36. All of the following are polar except
    A) S_3^{2-} B) NH_2^- C) I_3^- D) O_3 E) I_3^+
    Ans: C

37. All of the following are polar except
    A) SF_4 B) ClO_2^- C) IF_4^+ D) XeF_4 E) ClF_3
    Ans: D

38. Which of the following is polar?
    A) SF_6 B) ICl_4^- C) SF_4 D) AsF_6^- E) XeF_4
    Ans: C

39. All of the following are polar except
    A) ClF_3 B) $COCl_2$ C) CH_3^+ D) BrO_3^- E) O_3
    Ans: C

40. All of the following are polar except
    A) O_3 B) ClF_3 C) $COCl_2$ D) BrO_3^- E) CS_3^{2-}
    Ans: E

41. All of the following are polar except
    A) XeO_2 B) ClF_3 C) XeF_4 D) $SOCl_2$ E) XeO_3
    Ans: C

42. The molecule *cis*-dichloroethene is nonpolar. True or False?
    Ans: False

43. How many σ- and π-bonds, respectively, are there in acrolein, $CH_2=CHCHO$?
    A) 4 and 2 B) 7 and 2 C) 5 and 2 D) 5 and 4 E) 7 and 1
    Ans: B

44. How many σ- and π-bonds, respectively, are there in peroxyacetylnitrate, $CH_3C(O)O-ONO_2$?
    A) 9 and 2 B) 10 and 2 C) 10 and 1 D) 8 and 4 E) 8 and 2
    Ans: B

45. How many σ- and π-bonds are present in diazomethane, CH_2NN?
    Ans: 4 sigma and 2 pi

46. Draw the Lewis structure of formamide, NH_2CHO, and give the number of lone pairs of electrons, and the number of sigma and pi bonds.
    Ans: 3 lone pairs, 5 sigma bonds, and 1 pi bond

47. How many pi bonds are present in the cyanamide ion, $NCNH^-$?
    Ans: 2 pi bonds

48. Identify the hybrid orbitals used by the underlined atom in acetone, $CH_3\underline{C}OCH_3$.
    A) sp^3d B) sp^2 C) Pure p_z orbitals are used in bonding. D) sp^3 E) sp
    Ans: B

49. The hybrid orbitals used by the underlined atoms in $CH_3\underline{C}HCH\underline{C}N$, from left to right, respectively, are
    A) sp^3 and sp
    B) sp^2 and sp
    C) sp^2 and sp^3
    D) sp^2 and sp^2
    E) sp and sp^3
    Ans: B

50. The hybrid orbitals used by the underlined atoms in $CH_3\underline{C}H_2\underline{O}CH_2CH_3$, from left to right, respectively, are
    A) sp and sp B) sp^3 and sp C) sp^3 and sp^3 D) sp and sp^3 E) sp^2 and sp^3
    Ans: C

51. The hybrid orbitals used by the underlined atoms in $\underline{C}H_2CH\underline{C}HO$, from left to right, respectively, are
    A) sp^3 and sp^2 B) sp^2 and sp^2 C) sp^2 and sp D) sp and sp E) sp^3 and sp
    Ans: B

52. For the Lewis structure of the cyanamide ion that contains two double bonds, $\underline{N}=C=NH^-$, the hybrid orbitals used by the underlined nitrogen atom and the carbon atom, respectively, are
    A) sp^2 and sp^3 B) sp and sp C) sp^2 and sp^2 D) sp and sp^3 E) sp^2 and sp
    Ans: E

53. The HNC bond angle in formamide, H_2NCHO, is _____.
    Ans: about 107°; 109° is acceptable

54. All of the following are paramagnetic except
    A) O_2^+ B) O_2^- C) N_2^{2+} D) N_2^{2-} E) O_2
    Ans: C

55. Which of the following is diamagnetic?
    A) O_2^{2-} B) S_2 C) O_2^- D) O_2^+
    Ans: A

56. What is the ground-state electron configuration of O_2^-?
    Ans: $(\sigma_{2s})^2(\sigma_{2s}*)^2(\sigma_{2p})^2(\pi_{2p})^4(\pi_{2p}*)^2(\pi_{2p}*)^1$

57. Which of the following is paramagnetic?
    A) N_2 B) B_2 C) O_2^{2-} D) C_2^{2-} E) B_2^{2-}
    Ans: B

58. Which of the following would have the longest bond?
    $B_2, C_2, N_2, C_2^{2-}, N_2^{2-}$
    Ans: B_2, bond order = 1

59. The bond order of N_2^{2+} is
    A) 2.5   B) 1   C) 2   D) 1.5   E) 3
    Ans: C

60. The bond order of O_2^{2+} is
    A) 1   B) 2   C) 3   D) 2.5   E) 1.5
    Ans: D

61. Which of the following has the longest bond?
    A) N_2 B) NO^- C) N_2^{2+} D) N_2^{2-} E) O_2^{2-}
    Ans: E

62. Which of the following is paramagnetic?
    A) N_2 B) N_2^{2+} C) O_2^{2-} D) N_2^{2-} E) NO^+
    Ans: D

63. Which of the following species has the shortest bond length?
    A) NO^{2-} B) NO^{2+} C) NO^- D) NO E) NO^+
    Ans: E

64. Which of the following species has two unpaired electrons?
    A) OF^+ B) NO^+ C) CO^+ D) NF^+ E) CF^+
    Ans: A

65. Which of the following is a p-type semiconductor?
    A)   selenium doped with indium
    B)   silicon doped with arsenic
    C)   GaAs with arsenic in excess of gallium
    D)   germanium doped with arsenic
    E)   silicon doped with phosporus
    Ans: A

66. Which of the following is an n-type semiconductor?
    A)   silicon doped with phosphorus
    B)   silicon doped with boron
    C)   GaAs with gallium in excess of arsenic
    D)   selenium doped with indium
    E)   germanium doped with indium
    Ans: A

67. Germanium is a semiconductor. Which of the following should be added in small amounts to produce a p-type semiconductor?
    A) Bi   B) As   C) P   D) Sb   E) B
    Ans: E

68. Gallium is a semiconductor. Which of the following should be added in small amounts to produce a p-type semiconductor?
    A) Si   B) Sb   C) B   D) P   E) As
    Ans: C

69. How many lone pairs of electrons are there in the Lewis structure of azidocarbonamide, $H_2NC(O)NNC(O)NH_2$?
    A) 8   B) 12   C) 10   D) 16   E) 6
    Ans: A

70. What is the approximate **NNC** bond angle in azidocarbonamide, $H_2NC(O)NNC(O)NH_2$?
    A) 118°   B) 180°   C) 90°   D) 107°   E) 109°
    Ans: A

71. What is the hybridization of the bolded atoms **NNC,** from left to right, in azidocarbonamide, $H_2NC(O)NNC(O)NH_2$?
    A) sp^3, sp, sp^2 D)   sp, sp, sp^2
    B) sp^2, sp, sp^3 E) sp^2, sp^2, sp^2
    C) sp^2, sp, sp^2
    Ans: E

72. How many σ- and π-bonds, respectively, are there in the Lewis structure of azidocarbonamide, $H_2NC(O)NNC(O)NH_2$?
    A) 14 and 3   B) 15 and 3   C) 14 and 2   D) 8 and 3   E) 11 and 3
    Ans: E

73. Two Lewis structures can be written for diazomethane, where the arrangement of atoms is H_2**C-N-N**. The hybrid orbitals used by the bold atoms in these Lewis structures are
    A) sp^3 or sp^2, and sp.   B) sp^2 and sp.   C) sp^3 and sp.   D) sp^3 or sp^2, and sp^2.
    Ans: A

74. The fact that B_2 has two unpaired electrons means the $2p_\pi$ molecular orbitals have higher energy than the $2p_\sigma$ molecular orbitals. True or False?
    Ans: False

75. How many peaks would you predict for the photoelectron spectrum of water using 1) the molecular orbital model and 2) the VSEPR model?
    Ans: molecular orbital, 4; VSEPR, 2; the experimental result is 4 peaks

76. The OSO bond angle in the sulfite ion is _____ (*greater than, equal to, less than*) 109.5°.
    Ans: less than

77. An AX_3E_2 molecule has a trigonal planar shape. True or False?
    Ans: False

78. Consider the following molecules:
    (a) I_2
    (b) O_3
    (c) I_3^-
    (d) CS_2
    (e) CO
    Which of the molecules is polar?
    A) (b) and (e)   B) (b) and (c)   C) (c) and (e)   D) only (e)
    Ans: A

79. What is the bond order in the OH radical?
    Ans: 0.5

80. When two atoms are brought together along the *x*-axis, what is the number of σ bonds that can be formed by overlap of *p*-orbitals on each atom?
    A) 0   B) 1   C) 2   D) 3
    Ans: B

81. What hybrid orbitals are used by the N atoms in urea, H_2NCONH_2?
    A) sp B) sp^2 C) sp^3 D) dsp^3
    Ans: C

82. In the NO molecule, which atom makes the larger contribution to the lowest energy molecular orbital?
    Ans: O

83. For A_2, the LCAO-MO, $\psi = c_A\psi_A + c_B\psi_B$, has $c_A = c_B$. True or False?
    Ans: True

84. For HF, the LCAO-MO, $\psi = c_H\psi_H + c_F\psi_F$, has $c_H = c_F$. True or False?
    Ans: False

85. For peroxyacetylnitrate, $CH_3C(O)\mathbf{O}—ONO_2$, what hybrid orbitals are used by the oxygen atom in bold?
    A) dsp B) sp C) sp^2 D) sp^3
    Ans: D

Chapter 4: The Properties of Gases

1. All of the following elements are gases at room temperature and atmospheric pressure except
   A) Rn   B) F   C) Br   D) N   E) Cl
   Ans: C

2. The height of mercury in a barometer is 77.5 cm. What height would water reach in a water barometer? The density of mercury is 13.6 g·cm^{-3} and that of water 1.00 g·cm^{-3}. The temperature is constant.
   A) 175 cm   B) 55.9 cm   C) 760 cm   D) 5.70 cm   E) 1.05×10^3 cm
   Ans: E

3. The height of water in a water barometer is 883 cm at 20°C. The density of water at 20°C is 0.998 g·cm^{-3}. What is the pressure?
   A)   88.3 kPa   D) 8.64×10^3 Pa
   B) 8.81×10^3 Pa   E)   86.4 kPa
   C)   101 kPa
   Ans: E

4. A pressure of 2.50×10^5 kg·m^{-1}·s^{-2} corresponds to
   A) 1.00 atm   B) 760 mmHg   C) 253 kPa   D) 2.50×10^5 Torr   E) 2.50 Pa
   Ans: C

5. The pressure at 20,000 feet above sea level is about 400. mmHg. This corresponds to
   A)   526 Pa   D)   0.526 kPa
   B)   53.3 kg·m^{-1}·s^{-2} E)   53.3 kPa
   C)   0.526 Torr
   Ans: E

6. What is the pressure inside a system when the system-side column in an open mercury manometer is 75.0 mm lower than the atmosphere side when the atmospheric pressure is 735 mmHg?
   A) 820 kPa   B) 10.0 kPa   C) 88.0 kPa   D) 75 kPa   E) 108 kPa
   Ans: E

7. An oxygen tank kept at 20°C contains 28.0 moles of oxygen and the gauge reads 31.0 atm. After two weeks, the gauge reads 10.5 atm. How many moles of oxygen were used during the two-week period?
   A)   9.25 moles   D)   18.5 moles
   B)   7.5 moles   E)   9.48 moles
   C)   20.5 moles
   Ans: D

8. The value of the gas law constant, R, in units of $L \cdot kPa \cdot mol^{-1} \cdot K^{-1}$ is
   A) 62.4   B) 0.0821   C) 0.0752   D) 7.62   E) 8.31
   Ans: E

9. The value of the gas law constant, R, in units of $m^3 \cdot Pa \cdot mol^{-1} \cdot K^{-1}$ is
   A) 8.31   B) 0.0821   C) 7.62   D) 0.0752   E) 62.4
   Ans: A

10. A badly tuned automobile engine can release about 50 moles of carbon dioxide per hour. At 35°C, what volume of carbon dioxide is released in a six-hour period if the atmospheric pressure is 740 Torr?
    A) 7.9×10^5 L   B) 7.8×10^3 L   C) 1.3×10^3 L   D) 10 L   E) 1.3×10^5 L
    Ans: B

11. What volume is occupied by 1.00 kg of helium at 5.00°C at a pressure of 735 Torr?
    A) 5.97×10^5 L
    B) 5.90×10^3 L
    C) 2.95×10^3 L
    D) 1.06×10^2 L
    E) 5.60×10^3 L
    Ans: B

12. What volume is occupied by 500. g of fluorine gas at 5.00°C at a pressure of 735 Torr?
    A) 5.59 L   B) 11.2 L   C) 621 L   D) 295 L   E) 311 L
    Ans: E

13. A sample of gas with a volume of 750 mL exerts a pressure of 98.0 kPa at 30°C. What pressure will the sample exert when it is compressed to 250 mL and cooled to −25°C?
    A) 353 kPa   B) 241 kPa   C) 359 kPa   D) 39.9 kPa   E) 26.7 kPa
    Ans: B

14. If 2.00 moles of an ideal gas at STP is subjected to a new pressure of 24.7 kPa, the volume of the gas will become
    A) 8.20 L   B) 22.4 L   C) 11.2 L   D) 184 L   E) 91.9 L
    Ans: D

15. The number of molecules in 22.4 L of nitrogen at exactly 0°C and 3.00 atm is
    A) 6.02×10^{24}
    B) 1.81×10^{24}
    C) 1.81×10^{23}
    D) 2.01×10^{23}
    E) 18.0×10^{-23}
    Ans: B

16. Calculate the pressure of 4.00 g of nitrogen gas in a 1.00-L container at 20.0°C.
    A) 0.469 atm   B) 6.87 atm   C) 96.2 atm   D) 0.235 atm   E) 3.44 atm
    Ans: E

17. Calculate the mass of oxygen gas required to occupy a volume of 6.00 L at a pressure of 20.9 kPa and a temperature of 37.0°C.
    A) 1.56 g   B) 0.408 g   C) 0.779 g   D) 13.1 g   E) 0.0487 g
    Ans: A

18. How many atoms of argon occupy 1.00 mL if the temperature is 300.°C and the pressure is 1.00×10^{-3} Torr?
    A) 3.22×10^{13} D) 2.44×10^{19}
    B) 1.28×10^{16} E) 2.80×10^{-11}
    C) 1.68×10^{13}
    Ans: C

19. What is the density of xenon gas at 750 kPa and 25°C?
    A) 5.30 g·L^{-1} B) 0.392 g·L^{-1} C) 39.7 g·L^{-1} D) 79.5 g·L^{-1} E) 19.9 g·L^{-1}
    Ans: C

20. What is the density of helium gas at 750 kPa and 25°C?
    A)   2.42 g·L^{-1} D)   0.161 g·L^{-1}
    B)   1.21 g·L^{-1} E)   0.0120 g·L^{-1}
    C)   0.605 g·L^{-1}
    Ans: B

21. What is the density of nitrogen gas at 740 Torr and −12°C?
    A) 0.637 g·L^{-1} B) 1.17 g·L^{-1} C) 1.27 g·L^{-1} D) 2.55 g·L^{-1} E) 9.56 g·L^{-1}
    Ans: C

22. What is the density of chlorine gas at 850 Torr and −12°C?
    A) 27.8 g·L^{-1} B) 3.70 g·L^{-1} C) 3.39 g·L^{-1} D) 7.40 g·L^{-1} E) 1.85 g·L^{-1}
    Ans: B

23. Which gas is most dense at 1 atm and 25°C?
    A)   hydrogen cyanide   D)   nitrogen
    B)   hydrogen sulfide   E)   ammonia
    C)   carbon dioxide
    Ans: C

24. Which gas is most dense at 1 atm and 25°C?
    A)   hydrogen cyanide   D)   carbon monoxide
    B)   hydrogen sulfide   E)   nitrogen dioxide
    C)   nitrogen monoxide
    Ans: E

25. If 250.0 mL of a gas at STP weighs 2.00 g, what is the molar mass of the gas?
    A)   28.0 g·mol^{-1} D)   56.0 g·mol^{-1}
    B)   179 g·mol^{-1} E)   44.8 g·mol^{-1}
    C)   8.00 g·mol^{-1}
    Ans: B

26. What is the molar mass of a gas whose density is 3.55 g·L^{-1} at 110°C and a pressure of 736 Torr?
    A)   15.4 g·mol^{-1} D)   152 g·mol^{-1}
    B)   79.5 g·mol^{-1} E)   33.1 g·mol^{-1}
    C)   115 g·mol^{-1}
    Ans: C

27. A 0.479-g sample of nitrogen, oxygen or neon gas occupies a volume of 265 mL at 157 lPa and 20.0°C. What is the molar mass and identity of the gas?
    A)   14.0 g·mol^{-1}, N   D)   16.0 g·mol^{-1}, O
    B)   32.0 g·mol^{-1}, O_2 E)   28.0 g·mol^{-1}, N_2
    C)   20.1 g·mol^{-1}, Ne
    Ans: E

28. A sample of a gas weighing 15.1 g occupies 2.25 L at 1.75 atm and 20.0°C. If the empirical formula of the gas is NO_2, what is the molecular formula?
    A) N_5O_{10} B) N_3O_6 C) NO_2 D) N_4O_8 E) N_2O_4
    Ans: E

29. The empirical formula of a gas is CH_3O. If 2.77 g of the gas occupies 1.00 L at exactly 0°C at a pressure of 760 Torr, what is the molecular formula of the gas?
    A) $C_4H_{12}O_4$ B) $C_2H_6O_2$ C) $C_5H_{15}O_5$ D) CH_3O E) $C_3H_9O_3$
    Ans: B

30. Calculate the number of moles of oxygen gas collected by displacement of water at 14.0°C if the atmospheric pressure is 790. Torr and the volume is 5.00 L. The vapor pressure of water at 14.0°C is 12.0 Torr.
    A) 0.0184   B) 4.46   C) 0.00335   D) 0.217   E) 0.224
    Ans: D

31. A Group 17 or 18 gas has a density of 2.92 g·L^{-1} at 1.00 atm and 25°C. The gas is
    A) helium   B) neon   C) fluorine   D) argon   E) chlorine
    Ans: E

32. Calculate the density of camphor, $C_{10}H_{16}O$, at 80°C and 12 Torr.
    A) 8.2×10^{-4} g·L^{-1} D)   0.62 g·L^{-1}
    B)   0.083 g·L^{-1} E) 6.8×10^{-3} g·L^{-1}
    C)   0.37 g·L^{-1}
    Ans: B

33. At 80.0°C and 12.0 Torr, the density of camphor vapor is 0.0829 g·L^{-1}. What is the molar mass of camphor?
    A)   243 g·mol^{-1} D)   3490 g·mol^{-1}
    B)   34.5 g·mol^{-1} E)   20.3 g·mol^{-1}
    C)   152 g·mol^{-1}
    Ans: C

34. The density of the vapor of allicin, a component of garlic, is 1.14 g·L^{-1} at 125°C and 175 Torr. What is the molar mass of allicin?
    A)   21.6 g·mol^{-1} D)   869 g·mol^{-1}
    B)   50.8 g·mol^{-1} E)   162 g·mol^{-1}
    C)   273 g·mol^{-1}
    Ans: E

35. The density of citronellal, a mosquito repellant, is 1.45 g·L^{-1} at 365°C and 50.0 kPa. What is the molar mass of citronellal?
    A)   73.2 g·mol^{-1} D)   37.5 g·mol^{-1}
    B)   154 g·mol^{-1} E)   88.0 g·mol^{-1}
    C)   95.7 g·mol^{-1}
    Ans: B

36. Consider the following reaction:
    $$4KO_2(s) + 2CO_2(g) \rightarrow 2K_2CO_3(s) + 3O_2(g)$$
    How many moles of KO_2 are needed to react with 75.0 L of carbon dioxide at STP?
    A) 1.67   B) 6.70   C) 13.4   D) 0.838   E) 3.35
    Ans: B

37. Consider the following reaction:
    $$4KO_2(s) + 2CO_2(g) \rightarrow 2K_2CO_3(s) + 3O_2(g)$$
    How many moles of KO_2 are needed to react with 75.0 L of carbon dioxide at −25°C and 215 kPa?
    A) 3.91   B) 7.82   C) 31.3   D) 23.5   E) 15.6
    Ans: E

38. Consider the following reaction:
$$4KO_2(s) + 2CO_2(g) \rightarrow 2K_2CO_3(s) + 3O_2(g)$$
How many liters of oxygen are produced at STP if 10.5 moles of carbon dioxide are used at STP?
A) 15.8   B) 0.703   C) 706   D) 353   E) 235
Ans: D

39. Ammonium nitrate can decompose according to the following equation:
$$NH_4NO_3(s) \rightarrow N_2O(g) + 2H_2O(g)$$
How many liters of gas are produced by decomposition of 160 g of ammonium nitrate at STP?
A) 44.8   B) 6.00   C) 22.4   D) 134   E) 67.2
Ans: D

40. Ammonium nitrate can decompose according to the following equation:
$$NH_4NO_3(s) \rightarrow N_2O(g) + 2H_2O(g)$$
How many liters of gas are produced by decomposition of 160 g of ammonium nitrate at $-25°C$ and 86.0 kPa?
A) 48.0   B) 14.5   C) 57.6   D) 173   E) 144
Ans: E

41. What mass of aluminum metal is required to produce 2.24 L of hydrogen gas at exactly $0°C$ and 1.00 atm by reaction of the metal with excess strong acid?
A) 3.60 g   B) 4.05 g   C) 5.40 g   D) 2.70 g   E) 1.80 g
Ans: E

42. What mass of aluminum metal is required to produce 12.5 L of hydrogen gas at exactly $25°C$ and 1.00 atm?
$$2Al(s) + 6H^+(aq) \rightarrow 2Al^{3+}(aq) + 3H_2(g)$$
A) 20.7 g   B) 22.6 g   C) 10.0 g   D) 13.8 g   E) 9.19 g
Ans: E

43. Sulfur dioxide reacts with oxygen gas to produce sulfur trioxide. What volume of oxygen gas will react with 15.0 L of sulfur dioxide if both gases are at 101.3 kPa and $125°C$?
A) 30.0 L   B) 15.0 L   C) 7.50 L   D) 5.00 L   E) 3.75 L
Ans: C

44. Sulfur dioxide reacts with oxygen gas to produce sulfur trioxide. How many moles of oxygen gas will react with 15.0 L of sulfur dioxide if both gases are at 101.3 kPa and $125°C$?
A) 0.230   B) 0.115   C) 0.731   D) 0.919   E) 0.459
Ans: A

45. Calculate the mass of sulfur hexafluoride in a 2.00 L container at a pressure of 4.00 atm and a temperature of 78.0°C. The molar mass of sulfur hexafluoride is 146.1 g·mol^{-1}.
    A) 0.401 g   B) 40.6 g   C) 73.1 g   D) 36.6 g   E) 183 g
    Ans: B

46. Lithium metal reacts with nitrogen gas to produce lithium nitride. What volume of nitrogen gas at 2.00 atm and 175°C is required to produce 75.0 g of lithium nitride?
    A) 30.9 L   B) 119 L   C) 79.2 L   D) 39.6 L   E) 15.5 L
    Ans: D

47. Calcium cyanamide reacts with $H_2O(g)$ at 150°C and 1.00 atm pressure:
    $$CaNCN(s) + 3H_2O(g) \rightarrow CaCO_3(s) + 2NH_3(g)$$
    If 80.0 L of $NH_3(g)$ are produced, what volume of steam is required?
    A) 40.0 L   B) 80.0 L   C) 26.7 L   D) 120. L   E) 240. L
    Ans: D

48. A sample of nitrogen gas collected at 24°C and 745 Torr has a vapor pressure of 745 Torr. True or False?
    Ans: False

49. A mixture of oxygen and helium is 92.3% by mass oxygen. What is the partial pressure of oxygen if atmospheric pressure is 745 Torr?
    A) 412 Torr   B) 446 Torr   C) 688 Torr   D) 333 Torr   E) 299 Torr
    Ans: B

50. The main composition of dry air at sea level is about 75.5% nitrogen, 23.1% oxygen, and 1.3% argon. In a 1.00-g sample of dry air at 1.00 atm, calculate the partial pressure of argon gas.
    Ans: 9.6×10^{-3} atm

51. Oxygen can be produced in the laboratory by the following reaction:
    $$2KClO_3(s) \rightarrow 2KCl(s) + 3O_2(g)$$
    How many moles of potassium chlorate are needed to produce 257 mL of oxygen, collected over water at 14°C and 97.6 kPa? The vapor pressure of water at 14°C is 1.60 kPa.
    A) 6.89×10^{-3} D) 2.06×10^{-3}
    B) 7.00×10^{-3} E) 1.55×10^{-3}
    C) 7.12×10^{-3}
    Ans: A

52. Oxygen can be produced in the laboratory by the following reaction:
$$2KClO_3(s) \rightarrow 2KCl(s) + 3O_2(g)$$
How many moles of potassium chlorate are needed to produce 2.75 L of oxygen, collected over water at 37°C and 94.9 kPa? The vapor pressure of water at 37°C is 6.28 kPa.
A) 0.142   B) 6.75×10^{-2} C) 7.20×10^{-2} D) 6.30×10^{-2} E) 0.189
Ans: D

53. Which of the following gases effuses slowest?
   A)   fluorine
   B)   carbon dioxide
   C)   chlorine
   D)   carbon monoxide
   E)   nitrogen
   Ans: C

54. If it takes 15 s for a certain sample of neon to effuse through a porous barrier, it will take _____ (15 s, greater than 15 s, less than 15 s) for the same amount of nitrogen gas to effuse through the barrier under the same conditions.
   Ans: greater than 15 s

55. The following experiment was carried out using a newly synthesized chlorofluorocarbon. Exactly 50 mL of the gas effused through a porous barrier in 157 s. The same volume of argon effused in 76 s under the same conditions. Which compound is the chlorofluorocarbon?
   A) $C_2Cl_4F_2$ B) C_2ClF_5 C) $C_2Cl_2F_4$ D) C_2Cl_5F E) $C_2Cl_3F_3$
   Ans: C

56. The following experiment was carried out using a newly synthesized chlorofluorocarbon. Exactly 50 mL of the gas effused through a porous barrier in 172 s. The same volume of argon effused in 76 s under the same conditions. Which compound is the chlorofluorocarbon?
   A) $C_2Cl_2F_4$ B) $C_2Cl_4F_2$ C) $C_2Cl_3F_3$ D) C_2Cl_5F E) C_2ClF_5
   Ans: B

57. If the average speed of a water molecule at 25°C is 640 $m \cdot s^{-1}$, what is the average speed at 100°C?
   A) 572 $m \cdot s^{-1}$ B) 5120 $m \cdot s^{-1}$ C) 801 $m \cdot s^{-1}$ D) 320 $m \cdot s^{-1}$ E) 1280 $m \cdot s^{-1}$
   Ans: C

58. If the average speed of a carbon dioxide molecule is 410 $m \cdot s^{-1}$ at 25°C, what is the average speed of a molecule of methane at the same temperature?
   A) 680 $m \cdot s^{-1}$ B) 410 $m \cdot s^{-1}$ C) 1130 $m \cdot s^{-1}$ D) 1000 $m \cdot s^{-1}$ E) 247 $m \cdot s^{-1}$
   Ans: A

59. What is the root mean square speed of carbon dioxide molecules at 98°C?
    A) 153 m·s^{-1} B) 459 m·s^{-1} C) 45.6 m·s^{-1} D) 574 m·s^{-1} E) 236 m·s^{-1}
    Ans: B

60. Which of the following gases will have the largest root mean square speed at 100°C?
    A) water B) argon C) methane D) nitrogen E) oxygen
    Ans: C

61. A plot of the Maxwell distribution against speed for different molecules shows that
    A)   heavy molecules have a higher average speed.
    B)   light molecules have a very narrow range of speeds.
    C)   heavy molecules have a wide range of speeds.
    D)   light molecules have a lower average speed.
    E)   heavy molecules travel with speeds close to their average values.
    Ans: E

62. A plot of the Maxwell distribution for the same gas against temperature shows that
    A)   at high temperatures, most molecules have speeds close to their average speed.
    B)   as the temperature increases, a high proportion of molecules have very slow speeds.
    C)   as the temperature decreases, the spread of speeds widens.
    D)   as the temperature decreases, a high proportion of molecules have very high speeds.
    E)   at low temperatures, most molecules have speeds close to their average speed.
    Ans: E

63. Helium, H_2, and neon all have very small values of the van der Waals coefficient "a", whereas Cl_2, xenon, and water have large values. Which gases would you expect to be able to be liquefied by expansion?
    Ans: Cl_2, xenon, and water

64. Consider the following van der Waals coefficients:

Gas	a, $L^2 \cdot atm \cdot mol^{-2}$	b, $L \cdot mol^{-1}$
helium	0.034	0.0237
hydrogen	0.244	0.0266
neon	0.211	0.0171
krypton	2.32	0.0398
xenon	4.19	0.0511
chlorine	6.49	0.0562
carbon dioxide	3.59	0.0427
ammonia	4.17	0.0371
water	5.46	0.0305

Which of the following gases has the largest attractive forces?
A) neon   B) ammonia   C) chlorine   D) water   E) helium
Ans: C

65. Consider the following van der Waals coefficients:

Gas	a, $L^2 \cdot atm \cdot mol^{-2}$	b, $L \cdot mol^{-1}$
helium	0.034	0.0237
hydrogen	0.244	0.0266
neon	0.211	0.0171
krypton	2.32	0.0398
xenon	4.19	0.0511
chlorine	6.49	0.0562
carbon dioxide	3.59	0.0427
ammonia	4.17	0.0371
water	5.46	0.0305

Which of the following gases has the smallest attractive forces?
A) ammonia   B) hydrogen   C) neon   D) helium   E) chlorine
Ans: D

66. Which of the following gases would you predict to have the largest value of the van der Waals coefficient *b*?
A) $C_2F_2Cl_4$ B) CO_2 C) C_2F_6 D) Cl_2 E) C_2FCl_5
Ans: E

67. Which molecules of the following gases will have the greatest average kinetic energy?
A) CO_2 at 1 atm and 298 K.
B) N_2 at 1 atm and 298 K.
C)   All the molecules have the same kinetic energy.
D)   He at 0.1 atm and 298 K.
E) H_2 at 0.5 atm and 298 K.
Ans: C

68. Which molecules of the following gases will have the greatest root mean square speed?
    A)   nitrogen at 1 atm and 273 K.
    B)   All the molecules have the same root mean square speed.
    C)   hydrogen at 0.5 atm and 273 K.
    D)   argon at 1 atm and 273 K.
    E)   helium at 0.1 atm and 273 K.
    Ans: C

69. Consider two flasks at 25°C, one contains an ideal gas and the other contains the real gas SO_3. Which statement regarding these gases is true?
    A)   After the temperature is increased about 100 K, the pressure of the ideal gas will be smaller than the pressure of SO_3 because the van der Waals coefficient a for SO_3 is large.
    B)   As the temperature is decreased, the ideal gas will liquefy first because ideal gases have larger values of the van der Waals coefficient b.
    C)   After the temperature has been lowered about 100 K, the pressure of the ideal gas will be smaller than the pressure of SO_3 because the van der Waals coefficient a for SO_3 is large.
    D)   As the temperature approaches 0 K, the volume of the ideal gas will be larger than the volume of SO_3 because ideal gases lack intermolecular forces.
    E)   As the temperature approaches 0 K, the volume of the ideal gas will be smaller than the volume of SO_3 because ideal gases have larger values of the van der Waals coefficient a.
    Ans: D

70. Consider the following statements:
    1. Real gases act more like ideal gases as the temperature increases.
    2. When n and T are constant, a decrease in P results in a decrease in V.
    3. At 1 atm and 273 K, every molecule in a sample of a gas has the same speed.
    4. At constant T, CO_2 molecules at 1 atm and H_2 molecules at 5 atm have the same average kinetic energy.
    Which of these statements is true?
    A) 2 and 3   B) 1 and 2   C) 1 and 4   D) 3 and 4   E) 2 and 4
    Ans: C

71. Consider two cylinders of gas, both with a volume of 25 L and at 1 atm and 25°C. If one cylinder contains nitrogen gas and the other argon, which of the following statements are true and which are false.
    (a) The temperature of the gases is different.
    (b) The average molecular speed of the gases is the same.
    (c) The average kinetic energy of the gases is the same.
    Ans: (a) False
          (b) False
          (c) True

72. A 6.00-L sample of $C_2H_4(g)$ at 2.00 atm and 293 K is burned in 6.00 L of oxygen gas at the same temperature and pressure to form carbon dioxide and water. If the reaction goes to completion, what is the final volume of all gases at 2.00 atm and 293 K?
   A) 2.66 L   B) 6.00 L   C) 2.00 L   D) 1.33 L   E) 4.00 L
   Ans: A

73. A 1.00-L sample of $C_2H_4(g)$ at 2.00 atm and 293 K is burned in 8.00 L of oxygen gas at the same temperature and pressure to form carbon dioxide and water. If the reaction goes to completion, what is the final volume of all gases at 2.00 atm and 293 K?
   A) 5.33 L   B) 9.00 L   C) 8.00 L   D) 3.00 L   E) 5.00 L
   Ans: C

74. Real gases behave most nearly like ideal gases at
   A)   high temperatures and low pressures.
   B)   high pressures and low molar masses.
   C)   low temperatures and high pressures.
   D)   high temperatures and high pressures.
   E)   low temperatures and low pressures.
   Ans: A

75. Which of the following gases would not be expected to cool as it expands?
   A) NH_3 B) Cl_2 C) Xe   D) H_2O E) He
   Ans: E

76. What is the molar volume of a nitrogen gas at -27.0^oC and 1 atm?
   A) 2.22 L   B) 22.4 L   C) 24.6 L   D) 20.2 L
   Ans: D

77. What is the vapor pressure of geraniol, molar mass 155 g/mol, if the density of the vapor at 260°C is 0.480 g/L?
   A) 50.2 Torr   B) 103 Torr   C) 0.0661 Torr   D) 0.103 Torr
   Ans: B

78. How many liters of carbon dioxide measured at 25.0°C and 821 Torr would be produced by the combustion of 319 g of glucose, $C_6H_{12}O_6$?
   A) 3.37 L   B) 40.1 L   C) 241 L   D) 20.2 L
   Ans: C

79. How many liters of hydrogen gas measured at 25.0°C and 745 Torr are produced by detonation of 59.0 g of TNT, trinitrotoluene, $C_7H_5(NO_2)_3$?
   A) 0.544 L   B) 6.94 L   C) 1.36 L   D) 16.2 L
   Ans: D

80. How many liters of nitrogen gas measured at 25.0°C and 745 Torr are produced by detonation of 90.8 g of TNT, trinitrotoluene, $C_7H_5(NO_2)_3$?
    A) 0.839 L   B) 1.26 L   C) 9.98 L   D) 15.0 L
    Ans: D

81. The root mean square speed of nitrogen molecules in air at 20°C is 511 m/s in a certain container. If the gas is allowed to expand to twice its original volume, the root mean square velocity of nitrogen molecules drops to 325 m/s. Calculate the temperature after the gas has expanded.
    Ans: −154°C

82. If the compression factor, Z, is less than one for a given gas, the van der Waals coefficient "a" is dominant for real gases. True or False?
    Ans: True

83. For which of the following gases would you expect Z, the compression factor, to be greater than one at low pressures?
    A)   hydrogen
    B)   methane
    C)   ethane
    D)   All gases have Z less than one at low pressure.
    Ans: A

84. When referring to a gas, temperature is proportional to the square root of the average speed of the molecules. True or False?
    Ans: True

85. Consider both nitrogen and chlorine as ideal gases.
    (a) What is the molar kinetic energy of nitrogen gas at 25°C?
    (b) At 25°C, the molar kinetic energy of chlorine gas is
    _____(*equal to, greater than, less than*) the molar kinetic energy of nitrogen.
    Ans: (a) 3.72 kJ
         (b) equal to

Chapter 5: Liquids and Solids

1. Which of the following cations is likely to be hydrated in compounds?
   A) Rb^+ B) Li^+ C) K^+ D) Cs^+ E) NH_4^+
   Ans: B

2. Which of the following cations is likely to be hydrated in compounds?
   A) Rb^+ B) NH_4^+ C) Ba^{2+} D) K^+ E) Cs^+
   Ans: C

3. Which of the following cations is likely to be hydrated in compounds?
   A) Cs^+ B) NH_4^+ C) La^{3+} D) Rb^+ E) K^+
   Ans: C

4. Which of the following compounds is unlikely to be commonly available and thus found in the laboratory?
   A) $La(NO_3)_3 \cdot 6H_2O$ D) $CuSO_4 \cdot 5H_2O$
   B) $Na_2CO_3 \cdot 10H_2O$ E) $KCN \cdot 2H_2O$
   C) $BaCl_2 \cdot 2H_2O$
   Ans: E

5. All of the following hydrated compounds are commonly available except
   A) $BaCl_2 \cdot 2H_2O$ D) $Cr(ClO_4)_3 \cdot 6H_2O$
   B) $NH_4NO_3 \cdot 2H_2O$ E) $La_2(SO_4)_3 \cdot 9H_2O$
   C) $NaClO_4 \cdot H_2O$
   Ans: B

6. If the interaction between two species is proportional to $1/r^2$, which of the following is likely involved?
   A)   chloromethane molecules in the liquid phase
   B) Na^+ and H_2O
   C)   bromine molecules in the liquid phase
   D)   chloromethane molecules in the gas phase
   E)   ions in an ionic solid
   Ans: B

7. If the interaction between two species is proportional to $1/r^3$, which of the following is likely involved?
   A)   chloromethane molecules in the liquid phase
   B)   ions in an ionic solid
   C)   bromine molecules in the liquid phase
   D)   chloromethane molecules in the solid phase
   E) Na^+ and H_2O
   Ans: D

8. If the interaction between two species is proportional to $1/r^6$, which of the following is likely involved?
    A)   chloromethane molecules in the gas phase
    B)   chloromethane molecules in the solid phase
    C) Li^+ and H_2O
    D) Na^+ and H_2O
    E)   ions in an ionic solid
    Ans: A

9. Which will have the higher boiling point, 1,1-dichloroethene or *cis*-dichloroethene?
    Ans: 1,1-dichloroethene

10. Which of the following is true?
    A)   1,1-dichloro-2-methyl-1-propene has a higher boiling point than *trans*-2,3-dichloro-2-butene.
    B) CH_4 has a higher boiling point than CCl_4.
    C)   *o*-dichlorobenzene has a lower boiling point than *p*-dichlorobenzene.
    D)   HI has a lower boiling point than HBr.
    E)   Butane, C_4H_{10}, has a higher boiling point than acetone, CH_3COCH_3.
    Ans: A

11. Which of the following is true?
    A)   Butane, C_4H_{10}, has a higher boiling point than acetone, CH_3COCH_3.
    B)   Pentane, C_5H_{12}, has a lower boiling point than 2,2-dimethylpropane, C_5H_{12}.
    C) CHF_3 has a higher boiling point than CF_4.
    D) CH_4 has a higher boiling point than CCl_4.
    E)   HI has a lower boiling point than HBr.
    Ans: C

12. Which of the following has the highest boiling point?
    A) $CH_3CH_2CH_2CH_2CH_2Br$ D) $CH_3CH_2CH_2CH_2CH_2F$
    B) $CH_3CH_2CH_2CH_2CH_2I$ E) $CH_3CH_2CH_2CH_2CH_2Cl$
    C) $CH_3CH_2CH_2CH_2CH_3$
    Ans: B

13. Which of the following can form intermolecular hydrogen bonds?
    A) AsH_3 B) HBr   C) $CH_3CH_2CH_2CH_2CH_3$ D) CH_3OH E) CH_3OCH_3
    Ans: D

14. Which of the following can form intermolecular hydrogen bonds?
    A) $CH_3CH_2CH_2CH_2CH_3$ D) $CH_3CH_2NH_2$
    B) $CH_3CH_2C(O)H$ E) $CH_3CH_2OCH_2CH_3$
    C) $CH_3CH_2C(O)CH_3$
    Ans: D

15. Which of the following can form intermolecular hydrogen bonds?
    A) $(CH_3)_2NH$ B) PH_3 C) CH_3COCH_3 D) H_2CO
    Ans: A

16. Which of the following is the strongest intermolecular force between molecules?
    A)   dipole-dipole (stationary)   D)   London
    B)   hydrogen bonding   E)   ion-dipole
    C)   dipole-dipole (rotating)
    Ans: B

17. Which of the following has the highest boiling point?
    A) N_2 B) H_2S C) NH_3 D) H_2O E) SO_2
    Ans: D

18. Which will have the largest boiling point, HCl or HI?
    Ans: HI

19. Which of the following has the lowest boiling point?
    A) HF   B) GeH_4 C) SiH_4 D) PH_3 E) H_2Se
    Ans: C

20. What are all of the intermolecular forces that are responsible for the existence of ice?
    A)   dipole-dipole and London forces
    B)   London forces
    C)   dipole-dipole, London forces, and hydrogen bonding
    D)   dipole-dipole and ion-ion
    E)   hydrogen bonding and dipole-dipole
    Ans: C

21. What are all of the intermolecular forces that are responsible for the existence of the molecular solid oxalic acid, $H_2C_2O_4$?
    A)   dipole-dipole, London forces, and hydrogen bonding
    B)   dipole-dipole and London forces
    C)   hydrogen bonding and dipole-dipole
    D)   London forces
    E)   dipole-dipole and ion-ion
    Ans: A

22. Predict which of the following liquids has the lowest enthalpy of vaporization.
    A) H_2Se B) H_2O C) H_2Po D) H_2S E) H_2Te
    Ans: D

23. Predict which of the following has the lowest boiling point.
    A) SbH_3 B) AsH_3 C) NH_3 D) BiH_3 E) PH_3
    Ans: E

24. Which of the following molecular solids has the highest melting point?
    A)   sucrose, $C_{12}H_{22}O_{11}$ D) $TiCl_4$
    B)   carbon dioxide, CO_2 E)   benzene, C_6H_6
    C) H_2O
    Ans: A

25. Tetrabromomethane has a higher boiling point than tetrachloromethane.
    Ans: True

26. Pentane and 2,2-dimethylpropane have the formula C_5H_{12} and therefore have the same boiling point.
    Ans: False

27. Butane and 2-propanone have approximately the same boiling points.
    Ans: False

28. Given that $1\ N = 1\ kg{\cdot}m{\cdot}s^{-2}$, the units of surface tension are
    A) $kg{\cdot}s^{-2}{\cdot}m^2$ B) $N{\cdot}m$ C) $kg{\cdot}s^{-1}{\cdot}m^{-1}$ D) $N{\cdot}m^{-2}$ E) $kg{\cdot}s^{-2}$
    Ans: E

29. Water was found to rise to a height of 7.4 mm in a tube of internal radius 2.0 mm at room temperature. Given its density as $1.0\ g{\cdot}cm^{-3}$, what is the surface tension of water?
    A) $0.15\ N{\cdot}m^{-1}$ D) $0.073\ N{\cdot}m^{-1}$
    B) $0.036\ N{\cdot}m^{-1}$ E) $1.5 \times 10^{-7}\ N{\cdot}m^{-1}$
    C) $7.3 \times 10^{-8}\ N{\cdot}m^{-1}$
    Ans: D

30. Glycerol, $C_3H_8O_3$, has a higher viscosity than propanol, C_3H_8O.
    Ans: True

31. Viscosity usually increases with increasing temperature.
    Ans: False

32. The ability of water to "wet" paper, for example, is due to hydrogen bonding between water molecules and surface molecules in the paper.
    Ans: True

33. How many atoms are there in a face-centered cubic unit cell?
    Ans: 4

34. What is the coordination number of an atom in a primitive cubic structure?
    A) 6   B) 4   C) 8   D) 12
    Ans: A

35. The mass of a face-centered cubic unit cell is
    A)   two times the mass of one atom.   D)   six times the mass of one atom.
    B)   five times the mass of one atom.   E)   four times the mass of one atom.
    C)   equal to the mass of one atom.
    Ans: E

36. How many atoms are there in a primitive cubic unit cell?
    A) 1   B) 2   C) 3   D) 8   E) 4
    Ans: A

37. The coordination numbers in cubic close-packed, body-centered cubic, and primitive cubic structures are, respectively,
    A)   12, 8, and 12.   D)   12, 6, and 6.
    B)   12, 8, and 4.   E)   12, 8, and 6.
    C)   8, 12, and 6.
    Ans: E

38. How many tetrahedral and octahedral holes per atom are there in a cubic close-packed structure, respectively?
    A) 2 and 1   B) 0 and 1   C) 2 and 2   D) 2 and 0   E) 0 and 2
    Ans: A

39. Copper and silver are maleable because they have
    A)   coordination numbers of 12.   D)   hexagonal close-packed structures.
    B)   2 tetrahedral holes per atom.   E)   cubic close-packed structures.
    C)   primitive cubic structures.
    Ans: E

40. A cubic close-packed structure has
    A)   a coordination number of 4.
    B)   a coordination number of 6.
    C)   the same density as a hexagonal close-packed structure.
    D)   twice the density of a hexagonal close-packed structure.
    E)   a coordination number of 8.
    Ans: C

41. The atomic radius of magnesium is 160 pm. Estimate its density, given that the metal has a close-packed structure.
    A) 1.74 g·cm^{-3} D) 4.45 g·cm^{-3}
    B) 0.435 g·cm^{-3} E) 2.78 g·cm^{-3}
    C) 10.5 g·cm^{-3}
    Ans: A

42. The atomic radius of zinc is 137 pm. Estimate its density, given that the metal has a close-packed structure.
   A)   7.47 $g \cdot cm^{-3}$ D)   4.49 $g \cdot cm^{-3}$
   B)   10.2 $g \cdot cm^{-3}$ E)   19.2 $g \cdot cm^{-3}$
   C)   14.0 $g \cdot cm^{-3}$
   Ans: A

43. The atomic radius of aluminum is 143 pm. Estimate its density, given that the metal has a close-packed structure.
   A)   3.87 $g \cdot cm^{-3}$ D)   16.3 $g \cdot cm^{-3}$
   B)   5.54 $g \cdot cm^{-3}$ E)   2.71 $g \cdot cm^{-3}$
   C)   7.92 $g \cdot cm^{-3}$
   Ans: E

44. If the radius of an atom is r, what is the length of the side of the body-centered cubic unit cell?
   A) $4r/3^{\frac{1}{2}}$ B) 2.25r   C) r   D) 2r   E) $8^{\frac{1}{2}}r$
   Ans: A

45. If the radius of an atom is r, what is the length of the side of the face-centered cubic unit cell?
   A) $8^{\frac{1}{2}}r$ B) 2.25r   C) 2r   D) $4r/3^{\frac{1}{2}}$ E) r
   Ans: A

46. Estimate the density of cesium iodide from its crystal structure. The ionic radii of Cs^+ and Cl^- are 170 and 220 pm, respectively.
   A)   7.25 $g \cdot cm^{-3}$ D)   4.72 $g \cdot cm^{-3}$
   B)   18.9 $g \cdot cm^{-3}$ E)   3.77 $g \cdot cm^{-3}$
   C)   9.44 $g \cdot cm^{-3}$
   Ans: D

47. Estimate the density of magnesium oxide from its crystal structure. The radii of Mg^{2+} and O^{2-} are 72 and 140 pm, respectively.
   A)   3.51 $g \cdot cm^{-3}$ D)   1.76 $g \cdot cm^{-3}$
   B)   14.9 $g \cdot cm^{-3}$ E)   0.878 $g \cdot cm^{-3}$
   C)   21.1 $g \cdot cm^{-3}$
   Ans: A

48. Which answer best accounts for the Na^+ ions present in the NaCl unit cell?
    A) 1 center atom + 1/8 × 24 corner atoms
    B) 1 center atom + 1/2 × 6 face atoms
    C) 1 center + 1/8 × 8 corner atoms
    D) 1 center atom + 1/4 × 12 edge atoms
    E) 1/8 × 8 corner atoms + 1/2 × 6 face atoms
    Ans: D

49. Which answer best accounts for the Cl^- ions present in the NaCl unit cell?
    A) 1 center atom + 1/2 × 6 face atoms
    B) 1/8 × 8 corner atoms + 1/2 × 6 face atoms
    C) 1 center + 1/8 × 8 corner atoms
    D) 1 center atom + 1/4 × 12 edge atoms
    E) 1 center atom + 1/8 × 24 corner atoms
    Ans: B

50. The density of solid krypton is 2.16 g·cm^{-3}. If krypton crystallizes in a cubic close-packed structure, estimate the atomic radius.
    A) 66.5 pm   B) 225 pm   C) 113 pm   D) 318 pm   E) 450 pm
    Ans: B

51. The density of sodium metal is 0.97 g·cm^{-3}. If sodium crystallizes in a body-centered cubic structure, estimate the atomic radius.
    A) 95 pm   B) 320 pm   C) 370 pm   D) 65 pm   E) 190 pm
    Ans: E

52. Which of the following is anisotropic?
    A) an aqueous solution of sodium chloride
    B) a solid glass material
    C) a solid polyethylene plastic
    D) a sodium chloride crystal
    E) an aqueous solution of sugar
    Ans: D

53. Which of the following is isotropic?
    A) a sodium chloride crystal
    B) a cell membrane
    C) a solid polyethylene plastic
    D) a smectic liquid crystal
    E) a nematic liquid crystal
    Ans: C

54. All of the elements below can exist as network solids except
    A) O   B) Si   C) B   D) As   E) C
    Ans: A

55. How many calcium and fluorine ions are there in the fluorite unit cell shown below?

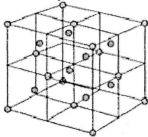

Ca (at opposite corners of small cubes)

F (at centers of small cubes)

A) $4\ Ca^{2+}$ and $2\ F^-$ D) $2\ Ca^{2+}$ and $8\ F^-$
B) $4\ Ca^{2+}$ and $8\ F^-$ E) $4\ Ca^{2+}$ and $4\ F^-$
C) $2\ Ca^{2+}$ and $4\ F^-$
Ans: B

56. What are the coordination numbers of Ca^{2+} and F^- ions, respectively, in fluorite? The unit cell is shown below

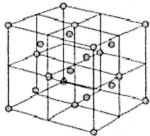

Ca (at opposite corners of small cubes)

F (at centers of small cubes)

A) 6 and 8   B) 4 and 8   C) 8 and 8   D) 6 and 4   E) 8 and 4
Ans: E

57. How many titanium and oxygen ions are there in the rutile unit cell shown below?

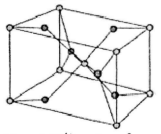

Ti

O

A) $3\ Ti^{4+}$ and $4\ O^{2-}$ D) $2\ Ti^{4+}$ and $3\ O^{2-}$
B) $3\ Ti^{4+}$ and $6\ O^{2-}$ E) $2\ Ti^{4+}$ and $6\ O^{2-}$
C) $2\ Ti^{4+}$ and $4\ O^{2-}$
Ans: C

58. What are the coordination numbers of Ti^{4+} and O^{2-}, respectively, in rutile? The unit cell is shown below.

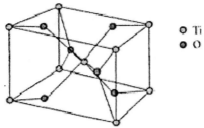

O Ti
● O

A) 4 and 3   B) 6 and 4   C) 3 and 6   D) 6 and 6   E) 6 and 3
Ans: E

59. How many calcium, titanium, and oxygen ions are there in the perovskite unit cell shown below?

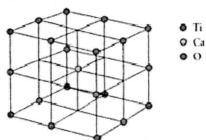

● Ti
○ Ca
● O

A)   1 Ca^{2+}, 2 Ti^{4+}, and 3 O^{2-} D)   1 Ca^{2+}, 1 Ti^{4+}, and 3 O^{2-}
B)   1 Ca^{2+}, 1 Ti^{4+}, and 6 O^{2-} E)   1 Ca^{2+}, 2 Ti^{4+}, and 6 O^{2-}
C)   1 Ca^{2+}, 4 Ti^{4+}, and 6 O^{2-}
Ans: D

60. What are the coordination numbers of calcium and titanium, respectively, in perovskite? The unit cell is shown below.

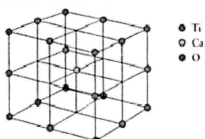

● Ti
○ Ca
● O

A) 6 and 6   B) 12 and 6   C) 12 and 8   D) 12 and 12   E) 6 and 8
Ans: B

61. What is the formula of the superconductor whose unit cell is given below? (The Y and Ba atoms are in the middle of the cell.)

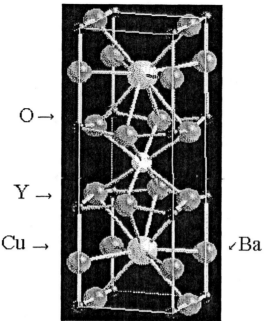

O →

Y →

Cu →   ⟍Ba

   A) $YBa_2Cu_3O_9$ D) $YBa_2Cu_4O_7$
   B) $YBa_2Cu_4O_5$ E) $YBa_2Cu_3O_7$
   C) $YBa_2Cu_2O_7$
   Ans: E

62. Which of the following statements is true?
   A)   A metallic conductor is a substance with a resistance that increases with increasing temperature.
   B)   A superconductor is a substance that has zero resistance below a certain temperature.
   C)   An insulator is a substance that does not conduct electricity below a certain temperature.
   D)   A semiconductor is a substance with a resistance that increases with increasing temperature.
   E)   An insulator behaves like a metallic conductor with a very high resistance.
   Ans: B

63. If the ratio of the radius of the cation to the anion in a 1:1 compound is 0.800,
   A)   the compound adopts the rock-salt structure.
   B)   the coordination number of the cation and anion are 6 and 6.
   C)   the compound adopts the same structure as calcium fluorite.
   D)   each cation occupies one-half of the tetrahedral holes of each anion cube.
   E)   each cation occupies all the cubic holes in the center of each anion cube.
   Ans: E

64. Both ZnS and CaF_2 have a face-centered cubic unit cell where the S^{2-} and Ca^{2+} ions are closest packed in each structure. Which of the following is true?
    A)   There are 4 tetrahedral holes empty in each structure.
    B)   In both compounds, one-half of the tetrhedral holes are filled.
    C)   In both compounds, all the tetrhedral holes are filled.
    D)   In ZnS, one-half of the tetrahedral holes are filled by Zn^{2+} ions whereas, in CaF_2, all the tetrahedral holes are filled with F^- ions.
    E)   There are 8 Zn^{2+} ions and 4 F^- ions in the unit cell.
    Ans: D

65. How many tetrahedral holes are there in a face-centered cubic unit cell?
    A) 8   B) 12   C) 1   D) 4   E) 6
    Ans: A

66. How many octahedral holes are there in a face-centered cubic unit cell?
    A) 6   B) 13   C) 12   D) 4   E) 8
    Ans: D

67. What is the coordination number of rubidium in RbF? The ionic radii of Rb^+ and F^- are 149 and 133 pm, respectively.
    A) 2   B) 4   C) 8   D) 6   E) 12
    Ans: C

68. What is the coordination number of magnesium in MgO? The ionic radii of Mg^{2+} and O^{2-} are 72 and 140 pm, respectively.
    A) 4   B) 2   C) 8   D) 6   E) 12
    Ans: D

69. Calculate the size of the octahedral holes in the NaCl structure. The ionic radii of Na^+ and Cl^- are 102 and 181 pm, respectively.
    A) 142 pm   B) 181 pm   C) 102 pm   D) 91 pm   E) 75 pm
    Ans: E

70. What is the coordination number of silver in AgCl? The ionic radii of Ag^+ and Cl^- are 113 and 181 pm, respectively.
    A) 4   B) 2   C) 6   D) 8   E) 12
    Ans: C

71. What is the coordination number of cesium in CsCl? The ionic radii of Cs^+ and Cl^- are 170 and 181 pm, respectively.
    A) 12   B) 8   C) 2   D) 4   E) 6
    Ans: B

72. Superconductivity is the loss of all electrical resistance when a substance is cooled below a certain characteristic transition temperature.
Ans: True

73. Metals with a hexagonal close-packed structure (have/do not have) slip planes and as a result are (malleable/brittle).
Ans: have; malleable

74. Which of the following are heterogeneous alloys?
A) mercury amalgam and tin-lead solder   D) tin-lead solder and bronze
B) brass   E) coinage cupronickel
C) bronze
Ans: A

75. The solid ZnS has a radius ratio of 0.45 and adopts the zinc-blend structure. What is the coordination number of Zn in ZnS?
A) 12   B) 4   C) 8   D) 2   E) 6
Ans: B

76. The boiling points of the Group 14 binary hydrides increase smoothly from CH_4 to SnH_4. True or False?
Ans: True

77. The boiling points of the Group 15 binary hydrides increase smoothly from NH_3 to SbH_3. True or False?
Ans: False

78. Which of the following liquids has the lowest viscosity?
A) acetone   B) ethanol   C) phosphoric acid   D) benzene
Ans: D

79. A certain liquid has a meniscus that curves downward in glass. This means that the cohesive forces in the liquid are less than the forces between the liquid and the glass. True or False?
Ans: False

80. An amorphous solid is one in which the atoms, ions, or molecules lie in a random jumble with no order. True or False?
Ans: True

81. In the zinc-blende structure (ZnS), the cations occupy half the tetrahedral holes. True or False?
Ans: True

82. In the rock-salt structure (NaCl), the cations occupy all the octahedral holes. True or False?
    Ans: True

83. Isotropic liquids have viscosities that are the same in every direction. True or False?
    Ans: True

84. In the unit cell of CaF_2, half of the tetrahedral holes are occupied by F^- ions. True or False?
    Ans: False

85. A new material will be recognized as a metal if
    A)   its electrical resistivity is independent of temperature.
    B)   its conductivity increases with increasing temperature.
    C)   its conductivity decreases with increasing temperature.
    D)   its electrical resistivity is zero.
    Ans: C

Chapter 6: Thermodynamics: The First Law

1. Hot coffee in a vacuum flask (thermos) is an example of a(n) _____
   (*open, closed, isolated*) system.
   Ans: isolated

2. A closed system can exchange energy with the surroundings. True or False?
   Ans: True

3. An isolated system can only exchange energy with the surroundings.
   Ans: False

4. Work is reported in joules and 1 joule = _____.
   Ans: $1 \text{ kg·m}^2\text{·s}^{-2}$

5. How much work is done by a person of mass 185 kg who climbs a ladder to the top of
   his house, a total of 15.0 m?
   A) 2.78 kJ   B) 12.3 kJ   C) 27.2 kJ   D) No work is done.   E) 121 kJ
   Ans: C

6. If an isolated system contained +5 kJ of energy, after 100 years ΔU =
   A)   impossible to determine   D)   0 kJ.
   B)   slightly less than +5 kJ.   E)   −5 kJ.
   C)   +5 kJ.
   Ans: C

7. What is the total motional contribution to the molar internal energy of gaseous HCN?
   A) 3RT   B) 3.5RT   C) 2.5RT   D) RT   E) 1.5RT
   Ans: C

8. What is the total motional contribution to the molar internal energy of gaseous BF_3?
   A) RT   B) 3.5RT   C) 3RT   D) 1.5RT   E) 2.5RT
   Ans: C

9. What is the total motional contribution to the molar internal energy of gaseous H_2O at
   25°C?
   A)   6.19 kJ·mol^{-1} D)   12.4 kJ·mol^{-1}
   B)   7.43 kJ·mol^{-1} E)   2.48 kJ·mol^{-1}
   C)   3.72 kJ·mol^{-1}
   Ans: B

10. A CD player and its battery together do 500 kJ of work, and the battery also releases 250 kJ of energy as heat and the CD player releases 50 kJ as heat due to friction from spinning. What is the change in internal energy of the system, with the system regarded as the battery and CD player together?
    A) +200 kJ   B) −700 kJ   C) −800 kJ   D) −200 kJ   E) −750 kJ
    Ans: C

11. A CD player and its battery together do 500 kJ of work, and the battery also releases 250 kJ of energy as heat, and the CD player releases 50 kJ as heat due to friction from spinning. What is the change in internal energy of the system, with the system regarded as the battery alone? Assume that the battery does 500 kJ of work on the CD player, which then does the same amount of work on the surroundings.
    A) +200 kJ   B) −800 kJ   C) −750 kJ   D) −50 kJ   E) −700 kJ
    Ans: C

12. A CD player and its battery together do 500 kJ of work, and the battery also releases 250 kJ of energy as heat and the CD player releases 50 kJ as heat due to friction from spinning. What is the change in internal energy of the system, with the system regarded as the CD player alone? Assume that the battery does 500 kJ of work on the CD player, which then does the same amount of work on the surroundings.
    A) −550 kJ   B) −50 kJ   C) −950 kJ   D) −800 kJ   E) +450 kJ
    Ans: B

13. When a gas expands into a vacuum, $w = 0$.
    Ans: True

14. A system had 150 kJ of work done on it and its internal energy increased by 60 kJ. How much energy did the system gain or lose as heat?
    A)   The system lost 90 kJ of energy as heat.
    B)   The system lost 210 kJ of energy as heat.
    C)   The system gained 60 kJ of energy as heat.
    D)   The system gained 90 kJ of energy as heat.
    E)   The system gained 210 kJ of energy as heat.
    Ans: A

15. If 2.00 mol of an ideal gas at 300 K and 3.00 atm expands from 6.00 L to 18.00 L and a final pressure of 1.20 atm, isothermally and reversibly, which of the following is correct?
    A) $w = -5.48$ kJ, $q = +5.48$ kJ, $\Delta U = 0$
    B) $w = -3.65$ kJ, $q = +3.65$ kJ, $\Delta U = 0$
    C) $w = +3.65$ kJ, $q = +3.65$ kJ, $\Delta U = +7.30$ kJ
    D) $w = -5.48$ kJ, $q = -5.48$ kJ, $\Delta U = -11.0$ kJ
    E) $w = +5.48$ kJ, $q = +5.48$ kJ, $\Delta U = +11.0$ kJ
    Ans: A

16. If 2.00 mol of an ideal gas at 300 K and 3.00 atm expands from 6.00 L to 18.00 L and a final pressure of 1.20 atm in two steps: (1) the gas is cooled at constant volume until its pressure has fallen to 1.20 atm, and (2) it is heated and allowed to expand against a constant pressure of 1.20 atm until its volume reaches 18.00 L, which of the following is correct?
    A) $w = 0$ for (1) and $w = -1.46$ kJ for (2)
    B) $w = -4.57$ kJ for the overall process
    C) $w = -6.03$ kJ for the overall process
    D) $w = -4.57$ kJ for (1) and $w = -1.46$ kJ for (2)
    E) $w = 0$ for the overall process
    Ans: A

17. When 2.00 kJ of energy is transferred as heat to nitrogen in a cylinder fitted with a piston at an external pressure of 2.00 atm, the nitrogen gas expands from 2.00 to 5.00 L against this constant pressure. What is ΔU for the process?
    A) −0.608 kJ   B) +2.61 kJ   C) −2.61 kJ   D) 0   E) +1.39 kJ
    Ans: E

18. A battery does 35 kJ of work driving an electric motor and 7 kJ of heat are released. What is the change in internal energy of the system?
    A) −35 kJ   B) +42 kJ   C) −42 kJ   D) −28 kJ   E) +28 kJ
    Ans: C

19. A piece of a newly synthesized material of mass 25.0 g at 80.0°C is placed in a calorimeter containing 100.0 g of water at 20.0°C. If the final temperature of the system is 24.0°C, what is the specific heat capacity of this material?
    A) 0.30 J·g^{-1}·(°C)$^{-1}$ D) 4.76 J·g^{-1}·(°C)$^{-1}$
    B) 7.46 J·g^{-1}·(°C)$^{-1}$ E) 0.84 J·g^{-1}·(°C)$^{-1}$
    C) 1.19 J·g^{-1}·(°C)$^{-1}$
    Ans: C

20. A piece of a newly synthesized material of mass 12.0 g at 88.0°C is placed in a calorimeter containing 100.0 g of water at 20.0°C. If the final temperature of the system is 24.0°C, what is the specific heat capacity of this material?
    A) 10.2 J·g^{-1}·(°C)$^{-1}$ D) 9.50 J·g^{-1}·(°C)$^{-1}$
    B) 1.58 J·g^{-1}·(°C)$^{-1}$ E) 0.54 J·g^{-1}·(°C)$^{-1}$
    C) 2.18 J·g^{-1}·(°C)$^{-1}$
    Ans: C

21. A reaction known to release 2.00 kJ of heat takes place in a calorimeter containing 0.200 L of solution and the temperature rose by 4.46°C. When 100 mL of nitric acid and 100 mL of sodium hydroxide were mixed in the same calorimeter, the temperature rose by 2.01°C. What is the heat output for the neutralization reaction?

    A)   0.0186 kJ·(°C)$^{-1}$ D)   17.9 kJ·(°C)$^{-1}$

    B)   0.448 kJ·(°C)$^{-1}$ E)   0.901 kJ·(°C)$^{-1}$

    C)   0.816 kJ·(°C)$^{-1}$

    Ans: E

22. A reaction known to release 4.00 kJ of heat takes place in a calorimeter containing 0.200 L of solution and the temperature rose by 6.14°C. When 200 mL of hydrochloric acid was added to a small piece of calcium carbonate in the same calorimeter, the temperature rose by 4.25°C. What is the heat output for this reaction?

    A)   0.0257 kJ·(°C)$^{-1}$ D)   2.77 kJ·(°C)$^{-1}$

    B)   0.941 kJ·(°C)$^{-1}$ E)   2.12 kJ·(°C)$^{-1}$

    C)   0.651 kJ·(°C)$^{-1}$

    Ans: D

23. All of the following are state functions except

    A) q   B) T   C) U   D) H

    Ans: A

24. Calculate the final temperature when 2.50 kJ of energy is transferred as heat to 1.50 mol N_2 at 298 K and 1 atm at constant volume.

    A) 80°C   B) 145°C   C) 159°C   D) 120°C   E) 105°C

    Ans: E

25. Calculate the final temperature when 2.50 kJ of energy is transferred as heat to 1.50 mol BF_3 at 298 K and 1 atm at constant volume.

    A) 100°C   B) 80°C   C) 67°C   D) 125°C   E) 92°C

    Ans: E

26. Calculate the change in internal energy when 2.50 kJ of energy is transferred as heat to 1.50 mol N_2 at 298 K and 1 atm at constant volume.

    A) −2.50 kJ   B) +2.50 kJ   C) +1.00 kJ   D) −1.50 kJ   E) +1.50 kJ

    Ans: B

27. Calculate the change in internal energy when 2.50 kJ of energy is transferred as heat to 1.50 mol BF_3 at 298 K and 1 atm at constant volume.

    A) +1.00 kJ   B) +2.50 kJ   C) −1.50 kJ   D) −2.50 kJ   E) +1.50 kJ

    Ans: B

28. Calculate the final temperature when 2.50 kJ of energy is transferred as heat to 1.50 mol N_2 at 298 K and 1 atm at constant pressure.
    A) 75°C   B) 159°C   C) 80°C   D) 82°C   E) 92°C
    Ans: D

29. Calculate the final temperature when 2.50 kJ of energy is transferred as heat to 1.50 mol BF_3 at 298 K and 1 atm at constant pressure.
    A) 82°C   B) 80°C   C) 159°C   D) 75°C   E) 92°C
    Ans: D

30. Calculate the change in internal energy when 2.50 kJ of energy is transferred as heat to 1.50 mol N_2 at 298 K and 1 atm at constant pressure.
    A) +2.14 kJ   B) −1.79 kJ   C) +2.50 kJ   D) −2.50 kJ   E) +1.79 kJ
    Ans: E

31. Calculate the change in internal energy when 2.50 kJ of energy is transferred as heat to 1.50 mol BF_3 at 298 K and 1 atm at constant pressure.
    A) +1.56 kJ   B) +2.50 kJ   C) 1.87 kJ   D) −2.50 kJ   E) −1.87 kJ
    Ans: C

32. Calculate the enthalpy change that occurs when 1.00 kg of acetone condenses at its boiling point (329.4 K). The standard enthalpy of vaporization of acetone is 29.1 kJ·mol^{-1}.
    A) −502 kJ   B) −29.1 kJ   C) −2.91 × 10^4 kJ   D) +502 kJ   E) +29.1 kJ
    Ans: A

33. Calculate the enthalpy change that occurs when 1 lb (454 g) of mercury freezes at its freezing point (234.3 K). The standard enthalpy of fusion of mercury is 2.29 kJ·mol^{-1}.
    A) −2.29 kJ   B) −1.04 × 10^3 kJ   C) +5.18 kJ   D) +2.29 kJ   E) −5.18 kJ
    Ans: E

34. Calculate the standard enthalpy of vaporization of liquid bromine if the standard enthalpy of sublimation of solid bromine is +40.1 kJ·mol^{-1} and the standard enthalpy of fusion of solid bromine is +10.6 kJ·mol^{-1}.
    A)   −50.7 kJ·mol^{-1} D)   +14.8 kJ·mol^{-1}
    B)   −29.5 kJ·mol^{-1} E)   +29.5 kJ·mol^{-1}
    C)   +50.7 kJ·mol^{-1}
    Ans: E

35. How much heat is required to vaporize 50.0 g of water if the initial temperature of the water is 25.0°C and the water is heated to its boiling point where it is converted to steam? The specific heat capacity of water is 4.18 $J \cdot (°C)^{-1} \cdot g^{-1}$ and the standard enthalpy of vaporization of water at its boiling point is 40.7 $kJ \cdot mol^{-1}$.
    A) 169 kJ   B) 64.2 kJ   C) 40.7 kJ   D) 193 kJ   E) 23.5 kJ
    Ans: D

36. When a solution of 1.691 g of silver nitrate is mixed with an excess of sodium chloride in a calorimeter of heat capacity 216 $J \cdot (°C)^{-1}$, the temperature rises 3.03°C. What is the reaction enthalpy?
    A)   −65.7 $kJ \cdot mol^{-1}$ D)   +65.7 $kJ \cdot mol^{-1}$
    B)   −0.654 $kJ \cdot mol^{-1}$ E)   +0.654 $kJ \cdot mol^{-1}$
    C)   +111 $kJ \cdot mol^{-1}$
    Ans: A

37. The combustion of one mole of octane, $C_8H_{18}(l)$, to produce carbon dioxide and liquid water has $\Delta H_r = -5471$ $kJ \cdot mol^{-1}$ at 298 K. What is the change in internal energy for this reaction?
    A)   −5460 $kJ \cdot mol^{-1}$ D)   −5471 $kJ \cdot mol^{-1}$
    B)   −5482 $kJ \cdot mol^{-1}$ E)   −5493 $kJ \cdot mol^{-1}$
    C)   −5449 $kJ \cdot mol^{-1}$
    Ans: A

38. The combustion of one mole of ethanol, $C_2H_5OH(l)$, to produce carbon dioxide and gaseous water has $\Delta H_r = -1235$ $kJ \cdot mol^{-1}$ at 298 K. What is the change in internal energy for this reaction?
    A)   −1237 $kJ \cdot mol^{-1}$ D)   −1247 $kJ \cdot mol^{-1}$
    B)   −1240 $kJ \cdot mol^{-1}$ E)   −1223 $kJ \cdot mol^{-1}$
    C)   −1230 $kJ \cdot mol^{-1}$
    Ans: B

39. The combustion of one mole of octane, $C_8H_{18}(l)$, in a bomb calorimeter released 5460 kJ of heat at 298 K. The products of the reaction are carbon dioxide and liquid water. What is the enthalpy change for this reaction?
    A)   +5471 $kJ \cdot mol^{-1}$ D)   −5449 $kJ \cdot mol^{-1}$
    B)   −5460 $kJ \cdot mol^{-1}$ E)   +5460 $kJ \cdot mol^{-1}$
    C)   −5471 $kJ \cdot mol^{-1}$
    Ans: C

40. The combustion of one mole of ethanol, $C_2H_5OH(l)$, in a bomb calorimeter released 1240 kJ of heat. The products of the reaction are carbon dioxide and liquid water. What is the enthalpy change for this reaction?

A) -1245 kJ·mol^{-1} D) $+1240$ kJ·mol^{-1}
B) -1235 kJ·mol^{-1} E) -1240 kJ·mol^{-1}
C) $+1235$ kJ·mol^{-1}

Ans: B

41. Calculate the standard reaction enthalpy for the oxidation of nitric oxide to nitrogen dioxide,

$$2NO(g) + O_2(g) \rightarrow 2NO_2(g)$$

from the standard reaction enthalpies of

$$N_2(g) + O_2 \rightarrow 2NO(g) \qquad \Delta H° = +180.5 \text{ kJ·mol}^{-1}$$
$$2NO_2(g) \rightarrow N_2(g) + 2O_2(g) \qquad \Delta H° = -66.4 \text{ kJ·mol}^{-1}$$

A) -114.1 kJ·mol^{-1} D) -294.6 kJ·mol^{-1}
B) -246.9 kJ·mol^{-1} E) $+246.9$ kJ·mol^{-1}
C) $+114.1$ kJ·mol^{-1}

Ans: A

42. Calculate the standard reaction enthalpy for the reaction

$$NO_2(g) \rightarrow NO(g) + O(g)$$

from the standard enthalpy of formation of ozone, $+142.7$ kJ·mol^{-1}, and from the following:

$$O_2(g) \rightarrow 2O(g) \qquad \Delta H° = +498.4 \text{ kJ·mol}^{-1}$$
$$NO(g) + O_3(g) \rightarrow NO_2(g) + O_2(g) \qquad \Delta H° = -200 \text{ kJ·mol}^{-1}$$

A) $+306$ kJ·mol^{-1} D) $+355$ kJ·mol^{-1}
B) $+555$ kJ·mol^{-1} E) $+592$ kJ·mol^{-1}
C) $+192$ kJ·mol^{-1}

Ans: A

43. Calculate the standard reaction enthalpy for the following reaction.

$$N_2H_4(l) + H_2(g) \rightarrow 2NH_3(g)$$

Given: $N_2H_4(l) + O_2(g) \rightarrow N_2(g) + 2H_2O(g) \qquad \Delta H° = -543 \text{ kJ·mol}^{-1}$
$\qquad 2H_2(g) + O_2(g) \rightarrow 2H_2O(g) \qquad \Delta H° = -484 \text{ kJ·mol}^{-1}$
$\qquad N_2(g) + 3H_2(g) \rightarrow 2NH_3(g) \qquad \Delta H° = -92.2 \text{ kJ·mol}^{-1}$

A) -935 kJ·mol^{-1} D) -59 kJ·mol^{-1}
B) -1119 kJ·mol^{-1} E) -243 kJ·mol^{-1}
C) -151 kJ·mol^{-1}

Ans: C

44. Calculate the standard reaction enthalpy for the following reaction.

$$CH_4(g) + H_2O(g) \rightarrow CO(g) + 3H_2(g)$$

Given: $2H_2(g) + CO(g) \rightarrow CH_3OH(l)$ $\Delta H° = -128.3$ kJ·mol^{-1}

         $2CH_4(g) + O_2(g) \rightarrow 2CH_3OH(l)$ $\Delta H° = -328.1$ kJ·mol^{-1}

         $2H_2(g) + O_2(g) \rightarrow 2H_2O(g)$ $\Delta H° = -483.6$ kJ·mol^{-1}

A)   +155.5 kJ·mol^{-1} D)   +42.0 kJ·mol^{-1}

B)   + 216 kJ·mol^{-1} E)   +206.1 kJ·mol^{-1}

C)   +412.1 kJ·mol^{-1}

Ans: E

45. What mass of propane, $C_3H_8(g)$, must be burned to supply 2580 kJ of heat? The standard enthalpy of combustion of propane at 298 K is -2220 kJ·mol^{-1}.

A) 25.6 g   B) 51.2 g   C) 102 g   D) 75.9 g   E) 37.9 g

Ans: B

46. What mass of ethanol, $C_2H_5OH(l)$, must be burned to supply 500 kJ of heat? The standard enthalpy of combustion of ethanol at 298 K is -1368 kJ·mol^{-1}.

A) 126 g   B) 2.74 g   C) 16.8 g   D) 10.9 g   E) 29.7 g

Ans: C

47. Calculate the standard enthalpy of combustion of ethanol at 298 K from standard enthalpy of formation data.

A) -957.0 kJ·mol^{-1} D) -1922 kJ·mol^{-1}

B) -1367 kJ·mol^{-1} E) -401.7 kJ·mol^{-1}

C) -687.5 kJ·mol^{-1}

Ans: B

48. Calculate the standard enthalpy of combustion of butane, $C_4H_{10}(g)$, at 298 K from standard enthalpy of formation data.

Ans: -2877.04 kJ·mol^{-1}

49. If the standard enthalpy of combustion of ethanol, $C_2H_5OH(l)$, at 298 K is -1368 kJ·mol^{-1}, calculate the standard enthalpy of formation of ethanol. The standard enthalpies of formation of carbon dioxide and liquid water are -393.51 and -285.83 kJ·mol^{-1}, respectively.

A)   +688.7 kJ·mol^{-1} D) -276.5 kJ·mol^{-1}

B) -344.3 kJ·mol^{-1} E) -688.7 kJ·mol^{-1}

C)   +276.5 kJ·mol^{-1}

Ans: D

50. If the standard enthalpy of combustion of octane, $C_8H_{18}(l)$, at 298 K is -5471 kJ·mol^{-1}, calculate the standard enthalpy of formation of octane. The standard enthalpies of formation of carbon dioxide and liquid water are -393.51 and -285.83 kJ·mol^{-1}, respectively.
    Ans: -249 kJ·mol^{-1}

51. The standard enthalpy of formation of ammonium perchlorate at 298 K is -295.31 kJ·mol^{-1}. Write the equation that corresponds to this value.
    Ans: $\frac{1}{2}N_2(g) + \frac{1}{2}Cl_2(g) + 2O_2(g) + 2H_2(g) \rightarrow NH_4ClO_4(s)$

52. Match the values with the correct enthalpy change.

Enthalpy	Value
$\Delta H_f°(C(g))$	-1300 kJ·mol^{-1}
$\Delta H_f°(C(s), \text{(graphite)})$	$+717$ kJ·mol^{-1}
$\Delta H_c°(C_2H_2(g))$	$+31$ kJ·mol^{-1}
$\Delta H_f°(Br_2(g))$	0

Ans:

$\Delta H_f°(C(g))$	$+717$ kJ·mol^{-1}
$\Delta H_f°(C(s), \text{(graphite)})$	0
$\Delta H_c°(C_2H_2(g))$	-1300 kJ·mol^{-1}
$\Delta H_f°(Br_2(g))$	$+31$ kJ·mol^{-1}

53. The standard enthalpy of formation of gaseous hydrogen atoms at 298 K is $+217.$ kJ·mol^{-1}. Write the equation that corresponds to this value.
    Ans: $\frac{1}{2}H_2(g) \rightarrow H(g)$

54. The lattice enthalpy of calcium bromide is the energy change for the reaction
    A) $CaBr_2(s) \rightarrow Ca(g) + 2Br(g)$ D) $CaBr_2(s) \rightarrow Ca(g) + Br_2(g)$
    B) $CaBr_2(s) \rightarrow Ca^{2+}(g) + 2Br^-(g)$ E) $Ca(g) + 2Br(g) \rightarrow CaBr_2(g)$
    C) $Ca(s) + Br_2(l) \rightarrow CaBr_2(s)$
    Ans: B

55. The lattice enthalpy of calcium oxide is the energy change for the reaction
    A) $CaO(s) \rightarrow Ca(g) + O(g)$ D) $CaO(s) \rightarrow Ca^{2+}(g) + O^{2-}(g)$
    B) $CaO(s) \rightarrow Ca(g) + \frac{1}{2}O_2(g)$ E) $Ca(s) + \frac{1}{2}O_2(g) \rightarrow CaO(s)$
    C) $Ca(g) + O(g) \rightarrow CaO(g)$
    Ans: D

56. The standard enthalpy of formation of NaCl(s) is −411 kJ/mol. In a Born-Haber cycle for the formation of NaCl(s), which enthalpy change(s) is/are endothermic?
    A)   the lattice enthalpy of NaCl(s)
    B)   the electron affinity of chlorine
    C)   the reverse of the lattice enthalpy of NaCl(s)
    D)   All of the enthalpy changes are endothermic except for the standard enthalpy of formation of NaCl(s).
    Ans: A

57. Calculate the lattice enthalpy of silver chloride from the following data.
    enthalpy of formation of Ag(g): +284 kJ·mol^{-1}
    first ionization energy of Ag(g): +731 kJ·mol^{-1}
    enthalpy of formation of Cl(g): +122 kJ·mol^{-1}
    electron affinity of Cl(g): +349 ($\Delta H = -349$) kJ·mol^{-1}
    enthalpy of formation of AgCl(s): −127 kJ·mol^{-1}
    A)   1613 kJ·mol^{-1} D)   1359 kJ·mol^{-1}
    B)   915 kJ·mol^{-1} E)   661 kJ·mol^{-1}
    C)   1037 kJ·mol^{-1}
    Ans: B

58. Calculate the lattice enthalpy of potassium fluoride from the following data.
    enthalpy of formation of K(g): +89 kJ·mol^{-1}
    first ionization energy of K(g): +418 kJ·mol^{-1}
    enthalpy of formation of F(g): +79 kJ·mol^{-1}
    electron affinity of F(g): +328 ($\Delta H = -328$) kJ·mol^{-1}
    enthalpy of formation of KF(s): −567 kJ·mol^{-1}
    A)   1481 kJ·mol^{-1} D)   497 kJ·mol^{-1}
    B)   825 kJ·mol^{-1} E)   347 kJ·mol^{-1}
    C)   904 kJ·mol^{-1}
    Ans: B

59. Calculate the lattice enthalpy of calcium oxide from the following data.
    enthalpy of formation of Ca(g): +178 kJ·mol^{-1}
    first ionization energy of Ca(g): +590 kJ·mol^{-1}
    second ionization energy of Ca(g): +1150 kJ·mol^{-1}
    enthalpy of formation of O(g): +249 kJ·mol^{-1}
    first electron affinity of O(g): +141 ($\Delta H = -141$) kJ·mol^{-1}
    second electron affinity of O(g): −844 ($\Delta H = +844$) kJ·mol^{-1}
    enthalpy of formation of CaO(s): −635 kJ·mol^{-1}
    A)   1817 kJ·mol^{-1} D)   3754 kJ·mol^{-1}
    B)   1391 kJ·mol^{-1} E)   3505 kJ·mol^{-1}
    C)   2235 kJ·mol^{-1}
    Ans: E

60. Calculate the lattice enthalpy of potassium chloride given the following enthalpy data.

$$K(s) \rightarrow K(g) \qquad\qquad +89 \text{ kJ·mol}^{-1}$$
$$K(g) \rightarrow K^+(g) + e^- \qquad +418 \text{ kJ·mol}^{-1}$$
$$\frac{1}{2}Cl_2(g) \rightarrow Cl(g) \qquad +122 \text{ kJ·mol}^{-1}$$
$$Cl(g) \ e^- \rightarrow Cl^-(g) \qquad -349 \text{ kJ·mol}^{-1}$$
$$K(s) + \frac{1}{2}Cl_2(g) \rightarrow KCl(s) \qquad -437 \text{ kJ·mol}^{-1}$$

Ans: +717 kJ·mol^{-1}

61. Calculate the standard enthalpy of formation of potassium chloride given the following enthalpy data.

$$K(s) \rightarrow K(g) \qquad\qquad +89 \text{ kJ·mol}^{-1}$$
$$K(g) \rightarrow K^+(g) + e^- \qquad +418 \text{ kJ·mol}^{-1}$$
$$\frac{1}{2}Cl_2(g) \rightarrow Cl(g) \qquad +122 \text{ kJ·mol}^{-1}$$
$$Cl(g) \ e^- \rightarrow Cl^-(g) \qquad -349 \text{ kJ·mol}^{-1}$$
$$KCl(s) \rightarrow K^+(g) + Cl^-(g) \qquad +717 \text{ kJ·mol}^{-1}$$

Ans: −437 kJ·mol^{-1}

62. The formation of solid calcium chloride from a gas of its ions is an endothermic process.
Ans: False

63. The formation of solid calcium oxide from a gas of its ions is an exothermic process and is equal to the reverse of the lattice enthalpy.
Ans: True

64. The standard enthalpy of formation of ammonia gas is −46.11 kJ·mol^{-1} at 298 K. What is the standard reaction enthalpy for the Haber process at 500°C? The molar heat capacities of nitrogen, hydrogen, and ammonia are 29.12, 28.82, and 35.06 J·K^{-1}·mol^{-1}.

A) −97.65 kJ·mol^{-1} D) −92.22 kJ·mol^{-1}
B) −113.81 kJ·mol^{-1} E) −103.09 kJ·mol^{-1}
C) −56.91 kJ·mol^{-1}
Ans: B

65. The standard enthalpy of formation of gaseous water is −241.82 kJ·mol^{-1} at 298 K. Estimate this value at 370 K. The molar heat capacities of hydrogen, oxygen, and gaseous water are 28.82, 29.36, and 33.58 J·K^{-1}·mol^{-1}.

A) −241.82 kJ·mol^{-1} D) −240.39 kJ·mol^{-1}
B) −243.25 kJ·mol^{-1} E) −241.10 kJ·mol^{-1}
C) −242.54 kJ·mol^{-1}
Ans: C

66. Calculate the average H-S bond enthalpy in $H_2S(g)$ given the standard enthalpies of formation for $H_2S(g)$, $H(g)$, and $S(g)$, -20.1, 218, and 223 kJ·mol^{-1}, respectively.
    A)   340 kJ·mol^{-1}
    B)   231 kJ·mol^{-1}
    C)   10.1 kJ·mol^{-1}
    D)   679 kJ·mol^{-1}
    E)   461 kJ·mol^{-1}
    Ans: A

67. Calculate the H-Br bond enthalpy given the standard enthalpies of formation for HBr(g), $H(g)$, and $Br(g)$, -36.2, 218, and 112 kJ·mol^{-1}, respectively.
    A)   320 kJ·mol^{-1}
    B)   366 kJ·mol^{-1}
    C)   124 kJ·mol^{-1}
    D)   284 kJ·mol^{-1}
    E)   196 kJ·mol^{-1}
    Ans: B

68. Calculate the Br-Br bond enthalpy given the standard enthalpies of formation for $Br_2(g)$ and $Br(g)$, 30.7 and 112 kJ·mol^{-1}, respectively.
    A)   193 kJ·mol^{-1}
    B)   255 kJ·mol^{-1}
    C)   143 kJ·mol^{-1}
    D)   81 kJ·mol^{-1}
    E)   30.7 kJ·mol^{-1}
    Ans: A

69. Calculate the standard enthalpy of formation of bicyco[1.1.0]butane (shown below)

    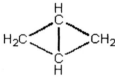

    given the standard enthalpies of formation of $C(g)$ and $H(g)$, 717 and 218 kJ·mol^{-1}, respectively, and the average C-H and C-C bond enthalpies, 412 and 348 kJ·mol^{-1}, respectively.
    A)   +175 kJ·mol^{-1}
    B)   −472 kJ·mol^{-1}
    C)   +312 kJ·mol^{-1}
    D)   −36 kJ·mol^{-1}
    E)   −124 kJ·mol^{-1}
    Ans: D

70. For the reaction $N_2O_5(s) \rightarrow 2NO_2(g) + \frac{1}{2}O_2(g)$, $\Delta H_r = 109.5$ kJ·mol^{-1} at 298 K. At constant temperature and pressure, which of the following statements is true?
    A) ΔH is independent of the physical states of the reactants.
    B) $\Delta H > \Delta U$
    C) $w = 0$
    D) $\Delta H = \Delta U$
    E) $\Delta H < \Delta U$
    Ans: B

71. For the reaction $CO_2(aq) \rightarrow CO_2(g)$, $\Delta H_r = 19.4$ kJ·mol^{-1} at 298. At constant temperature and pressure, which of the following statements is true?
    A) $w = 0$ B) $\Delta U = 2.48$ kJ   C) $\Delta H < \Delta U$ D) $\Delta H = \Delta U$ E) $\Delta H > \Delta U$
    Ans: E

72. The heat flow for the reaction
    $$C_2H_6(g) + 3.5O_2(g) \rightarrow 2CO_2(g) + 3H_2O(l)$$
    measured in a bomb calorimeter is -1553.5 kJ/mol at 298 K. At this temperature, ΔU is
    A) -1553.5 kJ/mol   C) -1547.3 kJ/mol
    B) -1552.3 kJ/mol   D) -1559.7 kJ/mol
    Ans: A

73. Give ΔH_r values for each of the following reactions from the information below.
    $\Delta H_f°(HCl(g)) = -92.31$ kJ·mol^{-1}
    H-Cl bond enthalpy $= +431$ kJ·mol^{-1}
    O-H bond enthalpy $= +463$ kJ·mol^{-1}
    $\Delta H_f°(H_2O(g)) = -241.8$ kJ·mol^{-1}
    (a) $2H(g) + O(g) \rightarrow H_2O(g)$
    (b) $H_2(g) + Cl_2(g) \rightarrow 2HCl(g)$
    (c) $H(g) + Cl(g) \rightarrow HCl(g)$
    Ans: (a) -926 kJ·mol^{-1}
          (b) -184.62 kJ·mol^{-1}
          (c) -431 kJ·mol^{-1}

74. Determine $\Delta H_f°(HCl(g))$ from the following data.
    H-Cl bond enthalpy $= +431$ kJ·mol^{-1}
    $\Delta H_f°(H(g)) = +217.9$ kJ·mol^{-1}
    $\Delta H_f°(Cl(g)) = +121.4$ kJ·mol^{-1}
    A) $+92$ kJ·mol^{-1} D) $+431$ kJ·mol^{-1}
    B) -261 kJ·mol^{-1} E) -92 kJ·mol^{-1}
    C) -431 kJ·mol^{-1}
    Ans: E

75. Use the following information to determine the standard enthalpy of formation of $NH_3(g)$.
    N-H bond enthalpy $= 390$ kJ·mol^{-1}
    $\Delta H_f°(H(g)) = 217.9$ kJ·mol^{-1}
    $\Delta H_f°(N(g)) = 472.6$ kJ·mol^{-1}
    A) -44 kJ·mol^{-1} D) -83 kJ·mol^{-1}
    B) -691 kJ·mol^{-1} E) -1170 kJ·mol^{-1}
    C) -516 kJ·mol^{-1}
    Ans: A

76. An isothermal change is one that occurs at a constant temperature. True or False?
Ans: True

77. Combustion reactions can be exothermic or endothermic. True or False?
Ans: False

78. For any isothermal process, $\Delta U > 0$ for an ideal gas. True or False?
Ans: False

79. The molar heat capacity of a monatomic ideal gas is independent of temperature and pressure. True or False?
Ans: True

80. The enthalpy of sublimation of a substance is related to its enthalpy of vaporization and enthalpy of fusion by the equation, $\Delta H_{sublimation} = \Delta H_{vaporization} - \Delta H_{fusion}$. True or False?
Ans: True

81. The temperature of a sample increases at its boiling point until all the liquid has been vaporized. True or False?
Ans: False

82. For most reactions, the value of ΔU is very close to that for ΔH. True or False?
Ans: True

83. The standard enthalpy of formation of $Br_2(l)$ is _____ (positive, negative, zero).
Ans: zero

84. All bond enthalpies are positive. True or False?
Ans: True

85. In the combustion of coal ($C(s)$), the C=O bonds are stronger than the O=O bond. True or False?
Ans: True

Chapter 7: Thermodynamics: The Second and Third Laws

1. Calculate the change in entropy of a large pail of water after 200 J of energy is reversibly transferred to the water at 20°C.
   A) -0.733 J·K^{-1}
   B) $+0.683 \text{ J·K}^{-1}$
   C) -0.683 J·K^{-1}
   D) $+0.733 \text{ J·K}^{-1}$
   E) -200 J·K^{-1}
   Ans: B

2. For a given transfer of energy, a greater change in disorder occurs when the temperature is high.
   Ans: False

3. Entropy is a state system.
   Ans: True

4. The temperature of 2.00 mol Ne(g) is increased from 25°C to 200°C at constant pressure. Calculate the change in the entropy of neon. Assume ideal behavior.
   A) $+7.68 \text{ J·K}^{-1}$
   B) $+19.2 \text{ J·K}^{-1}$
   C) -7.68 J·K^{-1}
   D) -19.2 J·K^{-1}
   E) $+9.60 \text{ J·K}^{-1}$
   Ans: B

5. Calculate the change in molar entropy when 2.00 moles of ozone are compressed isothermally to one quarter of its original volume. Treat ozone as an ideal gas.
   A) -23.1 J·K^{-1}
   B) -10.0 J·K^{-1}
   C) -1.39 J·K^{-1}
   D) $+10.0 \text{ J·K}^{-1}$
   E) $+23.1 \text{ J·K}^{-1}$
   Ans: A

6. Consider the following processes. (Treat all gases as ideal.)
   1. The pressure of one mole of oxygen gas is allowed to double isothermally.
   2. Carbon dioxide is allowed to expand isothermally to 10 times its original volume.
   3. The temperature of one mole of helium is increased 25°C at constant pressure.
   4. Nitrogen gas is compressed isothermally to one half its original volume.
   5. A glass of water loses 100 J of energy reversibly at 30°C.
   Which of these processes leads to an increase in entropy?
   A) 1 and 4   B) 5   C) 3 and 5   D) 2 and 3   E) 1 and 2
   Ans: D

7. Consider the following processes. (Treat all gases as ideal.)
   1. The pressure of one mole of oxygen gas is allowed to double isothermally.
   2. Carbon dioxide is allowed to expand isothermally to 10 times its original volume.
   3. The temperature of one mole of helium is increased 25°C at constant pressure.
   4. Nitrogen gas is compressed isothermally to one half its original volume.
   5. A glass of water loses 100 J of energy reversibly at 30°C.
   Which of these processes leads to a decrease in entropy?
   A) 1 and 2   B) 2   C) 3 and 4   D) 1, 4, and 5   E) 1 and 3
   Ans: D

8. Calculate the change in molar entropy when the pressure of argon is allowed to double isothermally. Assume ideal behavior.
   A) $+1.39 \, J \cdot K^{-1} \cdot mol^{-1}$
   B) $-1.39 \, J \cdot K^{-1} \cdot mol^{-1}$
   C) $-4.16 \, J \cdot K^{-1} \cdot mol^{-1}$
   D) $-5.76 \, J \cdot K^{-1} \cdot mol^{-1}$
   E) $+5.76 \, J \cdot K^{-1} \cdot mol^{-1}$
   Ans: D

9. Calculate the normal boiling point of chloroform given that the standard entropy and enthalpy of vaporization of chloroform is $+93.7 \, J \cdot K^{-1} \cdot mol^{-1}$ and $31.4 \, kJ \cdot mol^{-1}$.
   Ans: 335 K

10. What is the change in entropy when the pressure of an ideal gas is increased at constant temperature? Choose from +, 0, −.
    Ans: −

11. Calculate the standard entropy of condensation of chloroform at its boiling point, 335 K. The standard molar enthalpy of vaporization of chloroform at its boiling point is $31.4 \, kJ \cdot mol^{-1}$.
    A) $-31.3 \, kJ \cdot K^{-1} \cdot mol^{-1}$
    B) $+93.7 \, J \cdot K^{-1} \cdot mol^{-1}$
    C) $-93.7 \, J \cdot K^{-1} \cdot mol^{-1}$
    D) $+31.4 \, kJ \cdot K^{-1} \cdot mol^{-1}$
    E) $+506 \, J \cdot K^{-1} \cdot mol^{-1}$
    Ans: C

12. All entropies of fusion are negative. True or False?
    Ans: False

13. Calculate the standard entropy of vaporization of ethanol at its boiling point, 352 K. The standard molar enthalpy of vaporization of ethanol at its boiling point is $40.5 \, kJ \cdot mol^{-1}$.
    A) $+40.5 \, kJ \cdot K^{-1} \cdot mol^{-1}$
    B) $+115 \, J \cdot K^{-1} \cdot mol^{-1}$
    C) $-40.5 \, kJ \cdot K^{-1} \cdot mol^{-1}$
    D) $+513 \, J \cdot K^{-1} \cdot mol^{-1}$
    E) $-115 \, J \cdot K^{-1} \cdot mol^{-1}$
    Ans: B

14. Calculate the standard entropy of fusion of ethanol at its melting point, 159 K. The standard molar enthalpy of fusion of ethanol at its melting point is 5.02 kJ·mol^{-1}.
    A) -5.02 kJ·K^{-1}·mol^{-1}
    B) -31.6 J·K^{-1}·mol^{-1}
    C) $+5.02$ kJ·K^{-1}·mol^{-1}
    D) $+31.6$ J·K^{-1}·mol^{-1}
    E) -44.0 J·K^{-1}·mol^{-1}
    Ans: D

15. The enthalpy of fusion of H_2O(s) at its normal melting point is 6.01 kJ·mol^{-1}. The entropy change for freezing 1 mole of water at this temperature is
    A) $+20.2$ J·K^{-1}·mol^{-1}
    B) 0 J·K^{-1}·mol^{-1}
    C) -20.2 J·K^{-1}·mol^{-1}
    D) $+22.0$ J·K^{-1}·mol^{-1}
    E) -22.0 J·K^{-1}·mol^{-1}
    Ans: E

16. The change in molar entropy for vaporization of all liquids is about the same.
    Ans: True

17. Use Trouton's constant to estimate the enthalpy of vaporization of diethyl ether, which boils at 309 K.
    A) $+85$ kJ·mol^{-1}
    B) $+26$ kJ·mol^{-1}
    C) $+275$ kJ·mol^{-1}
    D) -85 kJ·mol^{-1}
    E) $+3.6$ kJ·mol^{-1}
    Ans: B

18. Use Trouton's constant to estimate the enthalpy of condensation of diethyl ether, which boils at 309 K.
    A) -26 kJ·mol^{-1}
    B) $+26$ kJ·mol^{-1}
    C) -275 kJ·mol^{-1}
    D) -85 kJ·mol^{-1}
    E) $+85$ kJ·mol^{-1}
    Ans: A

19. The molar entropy of silver at 298 K is equal to the area under the curve obtained by plotting (from T = 0 to T = 298 K)
    A) C_p versus T.
    B) $\ln C_p$ versus T.
    C) $\ln(C_p/T)$ versus T.
    D) C_p/T versus T.
    E) C_p versus 1/T.
    Ans: D

20. The molar heat capacity of Cu(s) at 1 atm pressure has been measured over a range of temperatures from close to 0 K to 400 K. Describe how you would obtain the standard molar entropy of Cu(s) at 298 K.
    Ans: Plot C_p/T versus T. The area under the curve from 0 K to 298 K is S° at 298 K.

21. Rank the standard molar entropy of the following from lowest to highest.
    1. $H_2O(l)$
    2. $H_2O(g)$
    3. $H_2O_2(l)$
    4. $H_2O_2(aq)$
    A) 4<1<3<2   B) 1<3<4<2   C) 3<1<4<2   D) 2<4<3<1   E) 1<3<2<4
    Ans: B

22. Which of the following has the lowest standard molar entropy?
    A) C(graphite)   B) $P_4(s)$ C) $S_8(s)$ D) $C_{60}(s)$ E) C(diamond)
    Ans: E

23. For He(g, 10 atm) → He(g, 1 atm), state whether the entropy change is positive or negative.
    Ans: positive

24. For $CO_2(g)$ → $CO_2(aq)$, state whether the entropy change is positive or negative.
    Ans: negative

25. For the following reaction,

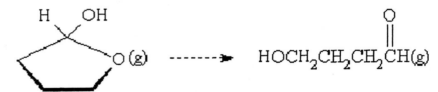

    state whether the entropy change is positive or negative.
    Ans: positive

26. The experimental value of the molar entropy of 1 mol NO at 0 K is about 5 $J·K^{-1}$. We can conclude that in the crystal the molecules of NO are arranged randomly. True or False?
    Ans: True

27. Use the Boltzmann formula to calculate the entropy at T = 0 of 1.00 mol chlorobenzene, C_6H_5Cl, where each molecule can be oriented in any of six ways.
    A)   0 $J·K^{-1}$; at T = 0, there is no randomness.
    B)   −15 $J·K^{-1}$
    C)   −30 $J·K^{-1}$
    D)   +30 $J·K^{-1}$
    E)   +15 $J·K^{-1}$
    Ans: E

28. Which of the following has the highest molar entropy?
    A) C(graphite)   B) $C_{60}(s)$ C) $CO_2(s)$ D) $CaCO_3(s)$ E) C(diamond)
    Ans: B

29. Which of the following has the smallest molar entropy?
    A) C(diamond)   B) $C_{60}(s)$ C) $CaCO_3(s)$ D) $CO_2(s)$ E) C(graphite)
    Ans: A

30. Which of the following has the smallest molar entropy at 298 K?
    A) $Cl_2(g)$ B) $N_2(g)$ C) He(g)   D) $F_2(g)$ E) Ne(g)
    Ans: C

31. Which of the following has the smallest entropy at 298 K?
    A) Kr(g)   B) $Br_2(l)$ C) Xe(g)   D) $Cl_2(g)$ E) $Br_2(g)$
    Ans: B

32. The molar entropy of lead at 298 K is equal to the area under the curve obtained by plotting (from T = 0 to 298 K)
    A) $\ln C_p$ versus T   D) $\ln(C_p/T)$ versus T
    B) C_p versus T   E) C_p versus 1/T
    C) C_p/T versus T
    Ans: C

33. Which of the following reactions has the largest value of $\Delta S°$?
    A) $NH_3(g) + HCl(g) \rightarrow NH_4Cl(s)$
    B) $2H_2(l) + O_2(l) \rightarrow 2H_2O(g)$
    C) $N_2(g) + 3H_2(g) \rightarrow 2NH_3(g)$
    D) $K(s) + O_2(g) \rightarrow KO_2(s)$
    E) $BaCl_2 \cdot 2H_2O(s) \rightarrow BaCl_2(s) + 2H_2O(g)$
    Ans: E

34. Which of the following would probably have a positive ΔS value?
    A) $He(g, 2 \text{ atm}) \rightarrow He(g, 10 \text{ atm})$ D) $O_2(g) \rightarrow O_2(aq)$
    B) $H_2(g) + I_2(s) \rightarrow 2HI(g)$ E) $2NO_2(g) \rightarrow N_2O_4(g)$
    C) $2Ag(s) + Br_2(l) \rightarrow 2AgBr(s)$
    Ans: B

35. Which of the following quantities is not equal to zero at 298 K?
    A) $\Delta H_f°(H^+(aq))$ D) $\Delta G_f°(H^+(aq))$
    B) $S°(H_2(g))$ E) $\Delta H_f°(H_2(g))$
    C) $S°(H^+(aq))$
    Ans: B

36. Sketch a plot of the molar entropy of oxygen gas from 0 K to 200 K. The normal melting and boiling points of oxygen are 55 K and 90 K, respectively.

    Ans: Key points: $S° = 0$ at 0 K; $S°$ increases with increasing temperature; vertical increases at 55 K and 90 K.

37. Calculate the standard entropy for the following reaction from standard molar entropies.

    $$NH_4ClO_4(s) + Al(s) \rightarrow NH_4Cl(s) + Al_2O_{3(s)}$$

    Ans: -69.01 J·K^{-1}·mol^{-1}

38. Which of the following is always true for a spontaneous process at constant temperature?

    A) $\Delta S_{system} + \Delta S_{surroundings} = q/T$ D) $\Delta S_{system} + \Delta S_{surroundings} > 0$

    B) $\Delta S > 0$ E) $\Delta S < q/T$

    C) $\Delta S = q/T$

    Ans: D

39. Calculate $\Delta S_{surr}°$ at 298 K for the reaction

        $H_2(g) + F_2(g) \rightarrow 2HF(g)$ $\Delta H_r° = -546$ kJ·mol^{-1}, $\Delta S_r° = +14.1$ J·K^{-1}·mol^{-1}

    A) $+14.1$ J·K^{-1}·mol^{-1} D) -1830 J·K^{-1}·mol^{-1}

    B) $+1820$ J·K^{-1}·mol^{-1} E) -14.1 J·K^{-1}·mol^{-1}

    C) $+1830$ J·K^{-1}·mol^{-1}

    Ans: C

40. Calculate $\Delta S_{surr}°$ at 298 K for the reaction

        $6C(s) + 3H_2(g) \rightarrow C_6H_6(l)$ $\Delta H_r° = +49.0$ kJ·mol^{-1}, $\Delta S_r° = -253$ J·K^{-1}·mol^{-1}

    A) $+164$ J·K^{-1}·mol^{-1} D) -164 J·K^{-1}·mol^{-1}

    B) -417 J·K^{-1}·mol^{-1} E) -253 J·K^{-1}·mol^{-1}

    C) $+253$ J·K^{-1}·mol^{-1}

    Ans: D

41. Calculate ΔS_{total} for the isothermal irreversible free expansion of 1.00 mol of ideal gas from 8.00 L to 20.00 L at 298 K.

    A) 0   D) -15.2 J·K^{-1}·mol^{-1}

    B) $+15.2$ J·K^{-1}·mol^{-1} E) -7.6 J·K^{-1}·mol^{-1}

    C) $+7.6$ J·K^{-1}·mol^{-1}

    Ans: C

42. The entropy of fusion of water is $+22.0$ J·K^{-1}·mol^{-1} and the enthalpy of fusion of water is $+6.01$ kJ·mol^{-1} at 0°C. At 0°C, ΔS_{total} for the melting of ice is

    A) -6010 J·K^{-1}·mol^{-1} D) $+6010$ J·K^{-1}·mol^{-1}

    B) 0   E) $+22.0$ J·K^{-1}·mol^{-1}

    C) -22.0 J·K^{-1}·mol^{-1}

    Ans: B

43. Consider the following reaction:
    $$2SO_2(g) + O_2(g) \rightarrow 2SO_3(g)$$
    Which statement is true for this reaction?
    A) $\Delta S < 0$ B) $\Delta S > 0$ C) $\Delta S = 0$ D) $S_m^{\circ} = 0$ for $O_2(g)$
    Ans: A

44. For the reaction
    $$2SO_3(g) \rightarrow 2SO_2(g) + O_2(g)$$
    $\Delta H_r^{\circ} = +198$ kJ·mol^{-1} at 298 K. Which statement is true for this reaction?
    A)   The reaction is driven by the enthalpy.
    B)   The reaction will not be spontaneous at low temperatures.
    C) ΔG_r° will be negative at high temperatures.
    D)   The reaction will not be spontaneous at any temperature.
    E) ΔG_r° will be positive at high temperatures.
    Ans: C

45. The reaction $2C(s) + 2H_2(g) \rightarrow C_2H_4(g)$ is endothermic. This reaction will not be spontaneous at any temperature.
    Ans: True

46. The reaction $N_2(g) + 3H_2(g) \rightarrow 2NH_3(g)$ is exothermic. This reaction will be spontaneous at all temperatures.
    Ans: False

47. Under what conditions (e.g., constant P, etc.) are the following relations true?
    (a) $\Delta G = \Delta H - T\Delta S$
    (b) $q = \quad H$
    Ans: (a) constant T
          (b) constant P, PV work only

48. The reaction $CH_3CH_2CH_2CH_3(g) \rightarrow CH_3CH(CH_3)_2(g)$, is exothermic. This reaction will be spontaneous at high temperatures.
    Ans: False

49. Consider the following reaction:
    $$Cl_2(g) \rightarrow 2Cl(g)$$
    Which of the following statements regarding this reaction are true?
    A)   The reaction is spontaneous at high temperatures.
    B)   The reaction is spontaneous at low temperatures.
    C)   The reaction is not spontaneous at any temperature.
    D)   The reaction is spontaneous at all temperatures.
    Ans: A

50. Which one of the following statements is true?
    A) *Labile* is a term that refers to the thermodynamic tendency of a substance to decompose.
    B) Spontaneous reactions always have $\Delta G_r° > 0$.
    C) Spontaneous reactions always have $\Delta H_r° < 0$.
    D) Spontaneous reactions always have $\Delta S_r° > 0$.
    E) A thermodynamically stable compound is a compound with a negative standard free energy of formation.
    Ans: E

51. Which one of the following statements is true?
    A) *Labile* is a term that refers to the thermodynamic tendency of a substance to decompose.
    B) A thermodynamically unstable compound is a compound with a positive standard free energy of formation.
    C) Spontaneous reactions always have $\Delta S_r° > 0$.
    D) Spontaneous reactions always have $\Delta G_r° > 0$.
    E) Spontaneous reactions always have $\Delta H_r° < 0$.
    Ans: B

52. Calculate $\Delta G_r°$ for the decomposition of mercury(II) oxide at 298 K.
    $$2HgO(s) \rightarrow 2Hg(l) + O_2(g)$$

$\Delta H_f°$, kJ·mol^{-1}	−90.83		
$S_m°$, J·K^{-1}·mol^{-1}	70.29	76.02	205.14

    A) −117.1 kJ·mol^{-1} D) +117.1 kJ·mol^{-1}
    B) +246.2 kJ·mol^{-1} E) −246.2 kJ·mol^{-1}
    C) −64.5 kJ·mol^{-1}
    Ans: D

53. Calculate the standard free energy of formation of mercury(II) oxide at 298 K given

	HgO(s)	Hg(l)	O$_2$(g)
$\Delta H_f°$, kJ·mol^{-1}	−90.83	–	–
$S_m°$, J·K^{-1}·mol^{-1}	70.29	76.02	205.14

    A) +58.5 kJ·mol^{-1} D) −123.1 kJ·mol^{-1}
    B) +117.1 kJ·mol^{-1} E) −117.1 kJ·mol^{-1}
    C) −58.5 kJ·mol^{-1}
    Ans: C

54. The standard free energy of formation of $CS_2(l)$ is 65.27 $kJ \cdot mol^{-1}$ at 298 K. This means that at 298 K
    A) $CS_2(l)$ will not spontaneously form $C(s) + 2S(s)$.
    B) $CS_2(l)$ is thermodynamically unstable.
    C) $CS_2(l)$ is thermodynamically stable.
    D)   No catalyst can be found to decompose $CS_2(l)$ into its elements.
    E) $CS_2(l)$ has a negative entropy.
    Ans: B

55. Consider the following compounds and their standard free energies of formation:

1	2	3	4	5
$C_6H_{12}(l)$ cyclohexane	$CH_3OH(l)$ methanol	$N_2H_4(l)$ hydrazine	$H_2O_2(l)$ hydrogen peroxide	$CS_2(l)$ carbon disulfide
+6.4 $kJ \cdot mol^{-1}$	−166 $kJ \cdot mol^{-1}$	+149 $kJ \cdot mol^{-1}$	−120 $kJ \cdot mol^{-1}$	+65.3 $kJ \cdot mol^{-1}$

    Which of these liquids is/are thermodynamically stable?
    A) 2 and 4   B) 2 and 3   C) 1, 3, and 5   D) 1   E) 3
    Ans: A

56. Consider the following compounds and their standard free energies of formation:

1	2	3	4	5
$C_6H_{12}(l)$ cyclohexane	$CH_3OH(l)$ methanol	$N_2H_4(l)$ hydrazine	$H_2O_2(l)$ hydrogen peroxide	$CS_2(l)$ carbon disulfide
+6.4 $kJ \cdot mol^{-1}$	−166 $kJ \cdot mol^{-1}$	+149 $kJ \cdot mol^{-1}$	−120 $kJ \cdot mol^{-1}$	+65.3 $kJ \cdot mol^{-1}$

    Which of these liquids is/are thermodynamically unstable?
    A) 2   B) 1, 3, and 5   C) 1 and 4   D) 2 and 3   E) 2 and 4
    Ans: B

57. Estimate the minimum temperature at which magnetite can be reduced to iron by graphite.
    $$Fe_3O_4(s) + 2C(s) \rightarrow 2CO_2(g) + 3Fe(s) \qquad \Delta S_r° = +351.44 \ J \cdot K^{-1} \cdot mol^{-1}$$
    The standard molar enthalpies of formation of magnetite and $CO_2(g)$ are −1118.4 and −393.51 $kJ \cdot mol^{-1}$, respectively.
    A)   670°C
    B)   Magnetite cannot be reduced by carbon at any temperature.
    C)   787°C
    D)   943°C
    E)   1790°C
    Ans: A

58. For the reaction
$$2C(s) + 2H_2(g) \rightarrow C_2H_4(g)$$
$\Delta H_r° = +52.3$ kJ·mol^{-1} and $\Delta S_r° = -53.07$ J·K^{-1}·mol^{-1} at 298 K. This reaction will be spontaneous at
A)   no temperature.
B)   all temperatures.
C)   temperatures below 985 K.
D)   temperatures above 985 K.
E)   temperatures below 1015 K.
Ans: A

59. For the reaction
$$2C(s) + 2H_2(g) \rightarrow C_2H_4(g)$$
$\Delta H_r° = +52.3$ kJ·mol^{-1} and $\Delta S_r° = -53.07$ J·K^{-1}·mol^{-1} at 298 K. The reverse reaction will be spontaneous at
A)   temperatures below 985 K.
B)   temperatures above 985 K.
C)   temperatures below 1015 K.
D)   all temperatures.
E)   no temperatures.
Ans: D

60. The following reaction is endothermic.
$$2Cu(s) + CO_2(g) \rightarrow 2CuO(s) + C(s)$$
Which of the following statements is true?
A)   The reaction is not spontaneous at any temperature.
B)   The reaction is spontaneous at all temperatures.
C)   It is impossible to determine if the reaction is spontaneous without calculations.
D)   The reaction will only be spontaneous at low temperatures.
E)   The reaction will only be spontaneous at high temperatures.
Ans: A

61. For the reaction
$$2SO_3(g) \rightarrow 2SO_2(g) + O_2(g)$$
$\Delta H_r° = +198$ kJ·mol^{-1} and $\Delta S_r° = 190$ J·K^{-1}·mol^{-1} at 298 K. The equilibrium constant for this reaction will be greater than one at
A)   all temperatures.
B)   temperatures above 1315 K.
C)   temperatures below 1042 K.
D)   no temperature.
E)   temperatures above 1042 K.
Ans: E

62. All of the following compounds become less stable with respect to their elements as the temperature is raised except
A) $PCl_5(g)$.   B) $C_6H_{12}(l)$.   C) $N_2H_4(l)$.   D) $CuO(s)$.   E) $HCN(g)$.
Ans: E

63. Consider the following compounds:
    $PCl_5(g)$, $HCN(g)$, $CuO(s)$, $NO(g)$, $NH_3(g)$, $SO_2(g)$.
    Which compound will have approximately the same stability with respect to its elements if the temperature is raised?
    A) $CuO(s)$ B) $NO(g)$ C) $PCl_5(g)$ D) $HCN(g)$ E) $NH_3(g)$
    Ans: B

64. Consider the following compounds:
    $PCl_5(g)$, $HCN(g)$, $CuO(s)$, $NO(g)$, $NH_3(g)$, $SO_2(g)$.
    Which compound will become more stable with respect to its elements if the temperature is raised?
    A) $CuO(s)$ B) $PCl_5(g)$ C) $NH_3(g)$ D) $NO(g)$ E) $HCN(g)$
    Ans: E

65. Consider the following compounds:
    $PCl_5(g)$, $HCN(g)$, $CuO(s)$, $NO(g)$, $NH_3(g)$, $SO_2(g)$.
    Which compound will become more stable with respect to its elements if the temperature is raised?
    A) $SO_2(g)$ B) $CuO(s)$ C) $NH_3(g)$ D) $PCl_5(g)$ E) $NO(g)$
    Ans: A

66. Which of the following statements is true?
    A)   If a reaction has $\Delta G_r° = -275$ kJ·mol^{-1}, it must proceed rapidly toward equilibrium.
    B)   All endothermic reactions are nonspontaneous.
    C)   The value of $\Delta G_r°$ is not dependent on temperature.
    D)   A spontaneous reaction for which the entropy change is negative is entropy driven.
    E)   If a certain reaction is spontaneous, it is not spontaneous in the reverse direction.
    Ans: E

67. Which of the following statements is true?
    A)   The molar entropy does not depend on the structure of the compound.
    B)   The molar entropy of $H_2O(l)$ is about the same as the molar entropy of ice.
    C)   An isothermal process that leads to a decrease in free energy is spontaneous.
    D)   If $\Delta G_r° > 0$, then the reaction is spontaneous.
    E)   All processes that give positive changes in energy are spontaneous.
    Ans: C

68. The sublimation of solid carbon dioxide is a spontaneous process. Predict the sign (+, −, or 0) of $\Delta G_r°$, $\Delta H_r°$, and $\Delta S_r°$, respectively.
    A) −, 0, +   B) −, −, −   C) −, +, +   D) 0, +, +   E) −, +, −
    Ans: C

69. Which of the following is true for a spontaneous process at constant temperature?
    A) $\Delta S = q/T$ B) $\Delta S > q_{rev}/T$ C) $\Delta S < q/T$ D) $\Delta S > q/T$
    Ans: D

70. Calculate ΔG for the following process.
    $$He(g, 1\ atm, 298\ K) \rightarrow He(g, 10\ atm, 298\ K)$$
    Ans: $5.70\ kJ \cdot mol^{-1}$

71. Use tabulated thermodynamic data to estimate the temperature at which the vapor pressure of benzene is 1.33 kPa. Hint: Assume the enthalpy and entropy of the reaction are independent of temperature.
    Ans: 256 K

72. For the reaction $2NH_3(g)$ b $3H_2(g) + N_2(g)$, $K_p = 1.47 \times 10^{-6}$ at 298 K. Estimate the temperature at which $K_p = 0.0100$.
    Ans: 390 K

73. Use tabulated thermodynamic data to calculate the concentration of $CO_2(aq)$ in equilibrium with an external pressure of 2.50 atm $CO_2(g)$ at 298 K.
    Ans: 0.092 M

74. An example of a spontaneous process having $\Delta H \sim 0$ is
    A)   1 L He(1 atm, 298 K) + 1 L Ar(1 atm, 298 K) $\rightarrow$ 2 L He/Ar mixture(1 atm, 298 K).
    B)   evaporation of water at 100°C and 1 atm.
    C)   a ball rolling from the top of a hill to the bottom of a valley.
    D)   precipitation of AgBr(s) from a solution of $Ag^+(aq)$ and $Br^-(aq)$.
    E)   freezing of water at −10°C.
    Ans: A

75. A piece of equipment must be designed to contain water at temperatures well above its normal boiling point. If the final piece of equipment will withstand an internal pressure of 10.0 atm, calculate the highest temperature at which the system can be safely operated.
    Ans: 169°C

76. When calculating the entropy change as a result of transferring heat reversibly to or from a system, the temperature must be constant. True or False?
    Ans: True

77. $\Delta U = 0$ for the isothermal expansion of an ideal gas and therefore $\Delta S = -w_{rev}/T = nR\ln(V_2/V_1)$. True or False?
    Ans: True

78. The entropy of fusion of a substance is always smaller than its entropy of vaporization. True or False?
Ans: True

79. Calculate the entropy of vaporization of water at 25°C and 1 bar. The molar heat capacities of the liquid and gas are 75 and 34 $J \cdot K^{-1} \cdot mol^{-1}$, respectively. The molar enthalpy of vaporization of water at its normal boiling point is 40.7 $kJ \cdot mol^{-1}$.
Ans: +118 $J \cdot K^{-1} \cdot mol^{-1}$

80. When calculating the entropy of vaporization of water at 25°C, the dominant contribution is
A) the molar heat capacity of liquid water.
B) heating the water from 25°C to 100°C.
C) cooling the water from 100°C to 25°C.
D) the entropy of vaporization of water at its normal boiling point.
Ans: D

81. All the halogens exist as diatomic molecules at room temperature and 1 bar. Under these conditions, which of the halogens, F to I, has the smallest molar entropy?
Ans: $I_2(s)$

82. The standard reaction free energy for the oxidation of glucose is −2879 $kJ \cdot mol^{-1}$. What is the maximum nonexpansion work obtainable from 125 g glucose, $C_6H_{12}O_6(s)$?
Ans: +2000 kJ

83. The dehydrogenation of cyclohexane, $C_6H_{12}(l)$, to form benzene, $C_6H_6(l)$, is not spontaneous. However, the hydrogenation of ethene to form ethane is spontaneous. Calculate the reaction free energy for the coupled reaction below.
$C_6H_{12}(l) + 3CH_2{=}CH_2(g) \rightarrow C_6H_6(l) + CH_3CH_3(g)$
Ans: −205 $kJ \cdot mol^{-1}$

84. The hydrolysis of ATP to ADP is spontaneous. However, the reaction of glucose with monohydrogen phosphate, the first step in the oxidation of glucose, is not spontaneous. Determine whether the coupled reactions are spontaneous.
Ans: Yes, the coupled reaction has a negative reaction free energy.

85. Reactions with positive values of ΔS_r° always become spontaneous at low temperatures. True or False?
Ans: False

Chapter 8: Physical Equilibria

1. Which of the following has the highest vapor pressure?
   A) H_2O
   B) CH_3CH_2OH
   C) $CH_3CH_2OCH_2CH_3$
   D) CH_3OH
   E) CH_3CH_2COOH
   Ans: C

2. In a closed vessel containing water the pressure is 18 Torr. If we add more water to the vessel, the pressure
   A) remains the same.   B) becomes larger.   C) becomes smaller.
   Ans: A

3. The vapor pressure of water above 40 mL of water in a 100 mL container is 23.8 Torr at 25°C. What is the vapor pressure of the water if the volume of the container is changed to 50 mL?
   A)   about 25 Torr
   B)   about 20 Torr
   C)   23.8 Torr
   D)   47.6 Torr
   E)   11.9 Torr
   Ans: C

4. The vapor pressure of *cis*-dibromoethene is lower than the vapor pressure of *trans*-dibromoethene due to
   A)   dipole-dipole forces.
   B)   London forces.
   C)   hydrogen bonding.
   D)   ion-ion forces.
   E)   ion-dipole forces.
   Ans: A

5. Which liquid do you expect to be the most volatile: CH_3CHO or CH_3OCH_3?
   Ans: CH_3CHO

6. Estimate the enthalpy of vaporization of CCl_4 given that at 25°C and 58°C its vapor pressure is 107 and 405 Torr, respectively. Assume that the enthalpy of vaporization is independent of the temperature.
   A)   486 $J \cdot mol^{-1}$
   B)   48.6 $kJ \cdot mol^{-1}$
   C)   142 $kJ \cdot mol^{-1}$
   D)   3.98 $kJ \cdot mol^{-1}$
   E)   33.1 $kJ \cdot mol^{-1}$
   Ans: E

7. Estimate the enthalpy of vaporization of water given that at 25°C and 35°C its vapor pressure is 23.8 and 42 Torr, respectively. Assume that the enthalpy of vaporization is independent of the temperature.

   A) 415 J·mol^{-1} D) 5.21 kJ·mol^{-1}
   B) 221 kJ·mol^{-1} E) 43.4 kJ·mol^{-1}
   C) 41.5 kJ·mol^{-1}
   Ans: E

8. What is the vapor pressure of carbon disulfide at its normal boiling point?
   Ans: 1 atm

9. The vapor pressure of water at 37°C is 47.1 Torr and its enthalpy of vaporization is 44.0 kJ·mol^{-1}. Estimate the vapor pressure of water at 87°C. Assume the enthalpy of vaporization of water is independent of temperature.
   A) 112 Torr   B) 256 Torr   C) 713 Torr   D) 52 Torr   E) 504 Torr
   Ans: E

10. A plot of ln(vapor pressure) versus 1/T for benzene gives a straight line with slope -3.70×10^3 K. The enthalpy of vaporization of benzene is
    A) 2.25 kJ·mol^{-1}.
    B) Not enough information given to calculate.
    C) 3.70 kJ·mol^{-1}.
    D) 30.8 kJ·mol^{-1}.
    E) 445 J·mol^{-1}.
    Ans: D

11. A plot of ln(vapor pressure) versus 1/T for methanol gives a straight line with an intercept of 13.6. The entropy of vaporization of methanol is
    A) 611 J·mol^{-1}·K^{-1}.
    B) 113 J·mol^{-1}·K^{-1}.
    C) 1640 J·mol^{-1}·K^{-1}.
    D) Not enough information given to calculate.
    E) 13.6 J·mol^{-1}·K^{-1}.
    Ans: B

12. A plot of ln(vapor pressure) versus 1/T for methanol gives a straight line with an intercept of 13.6. The enthalpy of vaporization of methanol is
    A) 113 kJ·mol^{-1}.
    B) 13.6 kJ·mol^{-1}.
    C) 611 kJ·mol^{-1}.
    D) 1.64 kJ·mol^{-1}.
    E) Not enough information given to calculate.
    Ans: E

13. The vapor pressure of benzene at 25°C is 94.6 Torr and its enthalpy of vaporization is 30.8 kJ·mol^{-1}. Estimate the normal boiling point of benzene. Assume the enthalpy of vaporization is independent of temperature.
    A)   640 K   D)   624 K
    B)   358 K   E)   Not enough information given.
    C)   470 K
    Ans: B

14. The vapor pressure of methanol at 25°C is 123 Torr and its enthalpy of vaporization is 35.3 kJ·mol^{-1}. Estimate the normal boiling point of methanol. Assume the enthalpy of vaporization is independent of temperature.
    A)   450 K   D)   373 K
    B)   342 K   E)   Not enough information given.
    C)   315 K
    Ans: B

15. In a pressure cooker, the boiling point is higher than the normal boiling point.
    Ans: True

16. In a pressure cooker, the boiling point of water is less than 100°C.
    Ans: False

17. Which of the following liquids freeze at a lower temperature when pressure is applied?
    A)   water   D)   methanol
    B)   acetic acid   E)   carbon tetrachloride
    C)   benzene
    Ans: A

18. The phase diagram for a pure compound is given below. The triple point occurs at

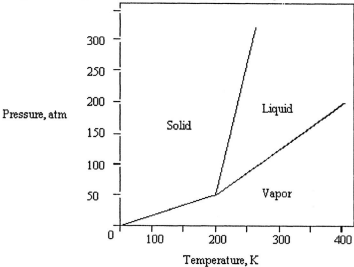

A)   50 atm and 200 K.
B)   0 atm and 200 K.
C)   greater than 50 atm and greater than 200 K.
D)   320 atm and 250 K.
E)   200 atm and 400 K.
Ans: A

19. The phase diagram for a pure compound is given below. All of the following could have a similar phase diagram except

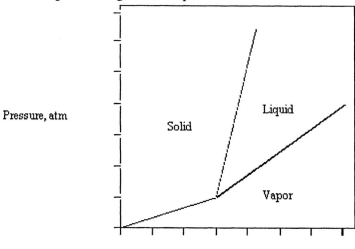

A)   methanol   D)   carbon tetrachloride
B)   carbon dioxide   E)   water
C)   benzene
Ans: E

20. The phase diagram for a pure substance is given below.

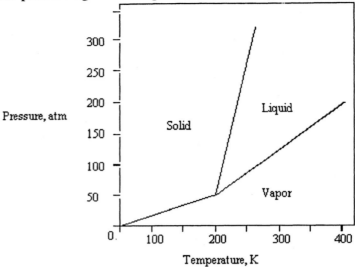

What is the lowest temperature at which liquid can exist?
A) 400 K   B) 0 K   C) 200 K   D) 150 K   E) 250 K
Ans: C

21. The phase diagram for a pure substance is given below. What is the critical temperature?

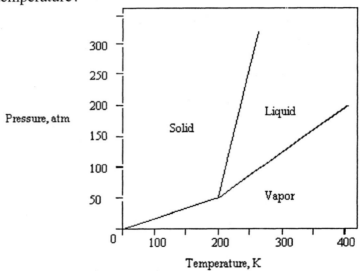

A) 0 K   B) 250 K   C) 300 K   D) 400 K   E) 200 K
Ans: D

22. The phase diagram for a pure substance is given below. What is the critical pressure?

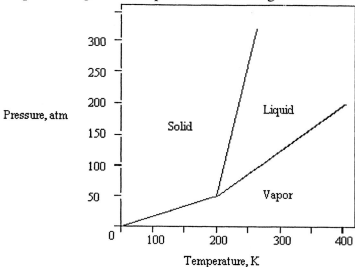

A)   50 atm
B)   150 atm
C)   325 atm
Ans: E

D)   any pressure above 325 atm
E)   200 atm

23. The phase diagram for a pure substance is shown below.

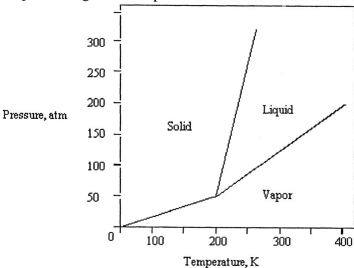

What is the highest temperature at which the substance can exist as a liquid?
A)   any temperature above 200 K
B)   250 K
C)   400 K
Ans: C

D)   200 K
E)   350 K

24. The phase diagram for a pure substance is given below.

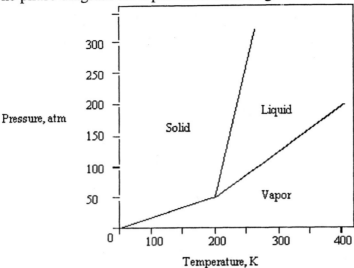

At 100 atm and 250 K, the substance exists as
A)   both vapor and liquid in equilibrium.   D)   vapor.
B)   liquid.   E)   solid.
C)   both vapor and solid in equilibrium.
Ans: B

25. The phase diagram for a pure substance is given below. The solid sublimes

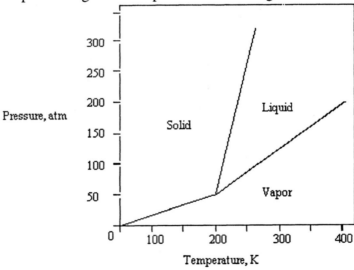

A)   at 400 K and 200 atm.
B)   at 200 K and 100 atm.
C)   at 300 K and 100 atm.
D)   at 300 K and 75 atm.
E)   if warmed at any pressure below 50 atm.
Ans: E

26. The phase diagram for a pure substance is given below.

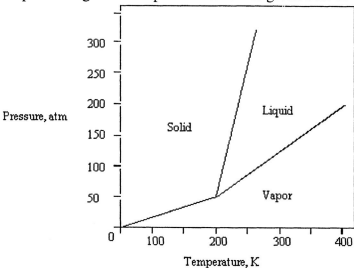

The substance is stored in a container at 150 atm at 25°C. Describe what happens if the container is opened at 25°C.

A)   The liquid in the container freezes.   D)   The vapor in the container escapes.
B)   The solid in the container sublimes.   E)   The liquid in the container vaporizes.
C)   The solid in the container melts.

Ans: E

27. The phase diagram for sulfur is given below.

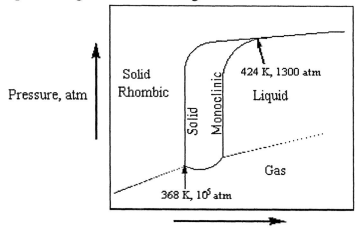

Which of the following is true?

A)   Sulfur has 2 triple points.
B)   Monoclinic sulfur does not sublime.
C)   Sulfur has 0 triple points.
D)   Sulfur has 3 triple points.
E)   Rhombic sulfur cannot be directly converted to liquid.

Ans: D

28. Consider the phase diagram for sulfur in the text. Monoclinic sulfur is denser than rhombic sulfur. True or False?

    Ans: False

29. The phase diagram for sulfur is given below.

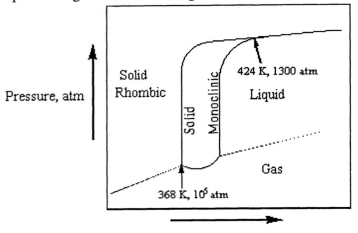

    At 424 K and 1300 atm,
    A) rhombic sulfur, monoclinic sulfur, sulfur liquid, and sulfur gas exist in equilibrium.
    B) only rhombic sulfur and sulfur gas exist in equilibrium.
    C) rhombic sulfur, monoclinic sulfur, and liquid sulfur exist in equilibrium.
    D) only rhombic sulfur is present.
    E) only monoclinic sulfur is present.

    Ans: C

30. The phase diagram for a pure substance is given below.

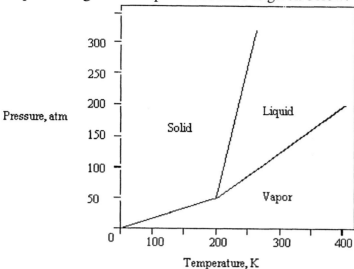

What pressure must be applied to liquefy a sample at 425 K?
A)   350 atm
B)   The sample cannot be liquefied at 425 K.
C)   150 atm
D)   50 atm
E)   250 atm
Ans: B

31. The phase diagram for sulfur is given below.

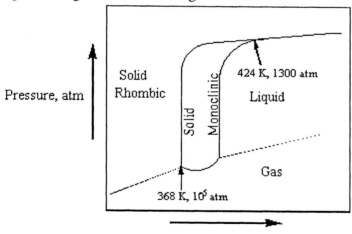

The number of phases that sulfur can exist in and the number of triple points, respectively, are
A) 3 and 2.   B) 3 and 3.   C) 4 and 2.   D) 3 and 1.   E) 4 and 3.
Ans: E

32. For a 1-component system, at the triple point
   A) $f = 3$.   B) $f = 2$.   C) $p = 1$.   D) $f = 0$.   E) $f = 1$.
   Ans: D

33. The phase diagram for carbon dioxide is given below.

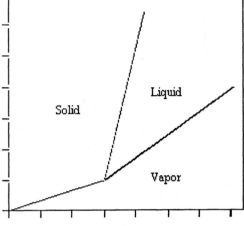

Temperature, K

If the triple point is at 5.1 atm and −56°C, at 1 atm and room temperature
   A)   solid carbon dioxide is the stable phase.
   B)   liquid carbon dioxide is the stable phase.
   C)   gaseous carbon dioxide condenses.
   D)   gaseous carbon dioxide is the stable phase.
   E)   solid carbon dioxide melts.
   Ans: D

34. The phase diagram for CO_2 is given below. The triple point is at 5.1 atm and 217 K.

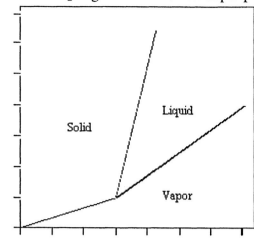

Pressure, atm

Solid

Liquid

Vapor

Temperature, K

What happens if $CO_2(l)$ at 30 atm and 450 K is released into a room at 1 atm and 298 K?
A)   The liquid vaporizes.
B)   The liquid remains stable.
C)   The liquid and vapor are in equilibrium.
D)   The liquid and solid are in equilibrium.
E)   The liquid freezes.
Ans: A

35. The phase diagram for CO_2 is given below. The triple point is at 5.1 atm and 217 K.

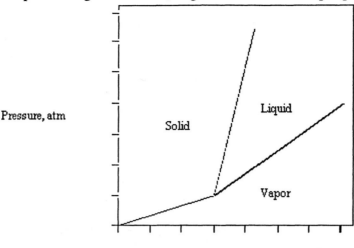

Temperature, K

What happens if carbon dioxide at −50°C and 25 atm is suddenly brought to 1 atm?
A)   The liquid and solid are in equilibrium.
B)   The solid melts.
C)   The solid and vapor are in equilibrium.
D)   The solid vaporizes.
E)   The solid remains stable.
Ans: D

36. The triple point for water is at 4.6 Torr and 0.01°C. The location of the triple point can vary depending on the conditions.
Ans: False

37. When three phases are in mutual equilibrium, such as at the triple point of water, the degrees of freedom equal 0.
Ans: True

38. Consider the phase diagrams for water and carbon dioxide given in the text. Explain the significance of the slopes of the solid-liquid lines.
Ans: For water, the solid is less dense than the liquid and the solid becomes less stable at high pressures. The opposite is true for carbon dioxide.

39. Consider the phase diagrams for water and carbon dioxide given in the text. Explain the following observations: A thin wire with weights attached is draped over a block of "dry ice" and a second wire with weights is draped over a block of ice. The wire cuts through the ice but not through the "dry ice."
Ans: For water, increased pressure reduces the melting point whereas for carbon dioxide the opposite is true. Remember, ice is less stable at high pressures.

40. A gas is a form of matter that can be liquefied at any temperature by pressure alone.
    True or False?
    Ans: False

41. Calculate the concentration of argon in lake water at 20°C. The partial pressure of argon
    is 0.0090 atm and Henry's constant is 0.0015 mol·L^{-1}·atm^{-1}.
    A) 1.5×10^{-3} M   D)   6.0 M
    B) 9.0×10^{-3} M   E)   0.17 M
    C) 1.4×10^{-5} M
    Ans: C

42. Which of the following would likely dissolve in benzene?
    A) NaCl   B) Cl_2CCCl_2 C) Na_2CO_3 D) $Ca(HCO_3)_2$ E) $C_6H_{12}O_6$
    Ans: B

43. Calculate the number of moles of oxygen that will dissolve in 45 L of water at 20°C
    if the partial pressure of oxygen is 0.21 atm. Henry's constant for oxygen is
    0.0013 mol·L^{-1}·atm^{-1}.
    A) 0.0062 M   B) 0.0013 M   C) 0.012 M   D) 0.00027 M   E) 0.28
    Ans: C

44. The "bends" result from the rapid release of nitrogen gas in the bloodstream when the
    diver returns to the surface. Would argon gas or hydrogen gas be suitable replacements
    for nitrogen gas in "compressed air?" Explain.
    Ans: No, both are more soluble in plasma than nitrogen.

45. The enthalpy of hydration of AgBr is −819 kJ·mol^{-1} at 25°C. If the hydration enthalpy
    of Br$^-$ is −309 kJ·mol^{-1}, calculate the enthalpy of hydration of Ag$^+$ ions.
    A)   −410 kJ·mol^{-1} D)   −309 kJ·mol^{-1}
    B)   −819 kJ·mol^{-1} E)   −510 kJ·mol^{-1}
    C)   −1128 kJ·mol^{-1}
    Ans: E

46. For AlF$_3$, the lattice enthalpy is 5220 kJ·mol^{-1} and the enthalpy of solution is
    −27 kJ·mol^{-1} at 25°C. Calculate the enthalpy of hydration of AlF$_3$ at this temperature.
    Ans: −5247 kJ·mol^{-1}

47. For CaCl$_2$, the enthalpies of hydration and solution are −2337 and −81 kJ·mol^{-1},
    respectively, at 25°C. Calculate the lattice enthalpy of calcium chloride.
    Ans: +2256 kJ·mol^{-1}

48. For AgI, the lattice enthalpy is larger than the absolute value of the enthalpy of hydration. This means that for AgI
    A) ΔH_{sol} is positive.
    B)   the solubility increases when the temperature decreases.
    C) ΔH_{hyd} is positive.
    D) ΔH_{sol} is negative.
    E) ΔH_L is negative.
    Ans: A

49. For $CaCl_2$, the absolute value of the enthalpy of hydration is larger than the lattice enthalpy. This means that for $CaCl_2$
    A)   the lattice enthalpy is negative.
    B)   the enthalpy of solution is endothermic.
    C)   the solubility increases when the temperature increases.
    D)   the enthalpy of hydration is positive.
    E)   the enthalpy of solution is exothermic.
    Ans: E

50. Which of the following is likely to have the largest exothermic hydration enthalpy?
    A) Sr^{2+} B) Na^+ C) Mg^{2+} D) Ca^{2+} E) Li^+
    Ans: C

51. Which of the following is likely to have the smallest exothermic hydration enthalpy?
    A) F^- B) I^- C) NO_3^- D) Cl^- E) Br^-
    Ans: C

52. Which of the following statements is true regarding solubility? Assume the solute does not affect the solvent and we are dealing with dilute solutions.
    A)   We can expect substances with positive enthalpies of solution always to be soluble.
    B)   We can expect substances with positive enthalpies of solution always to be soluble only if the entropy of solution is negative.
    C)   All the statements are false because the solute always affects the solvent resulting in a negative entropy of solution.
    D)   We can expect substances with negative enthalpies of solution always to be soluble.
    E)   We can expect substances with positive enthalpies of solution always to be soluble only if the entropy of solution is positive.
    Ans: D

53. What is the molality of sodium perchlorate, $NaClO_4$, in a solution preprated by dissolving 125 g sodium perchlorate monohydrate in 500.0 g water?
    A) 3.56 m   B) 1.78 m   C) 0.900 m   D) 2.04 m   E) 0.250 m
    Ans: B

54. What is the molality of CrCl₃ in a solution prepared by dissolving 75.2 g chromium(III) chloride hexahydrate in 250.0 g of water.
A) 0.282 m   B) 1.13 m   C) 7.60 m   D) 5.64 m   E) 1.90 m
Ans: B

55. If you wanted to study solutions of different concentrations at different temperatures, would you use molarity or molality as your measure of concentration? Why?
Ans: Molality because it is temperature independent.

56. Calculate the molality of ethanol in a solution of water, given that the mole fraction of ethanol is 0.300.
Ans: 23.8 m

57. What is the molality of carbon tetrachloride in a solution of toluene, given that the mole fraction is 0.200?
Ans: 2.71 m

58. Aqueous ammonia (28%) is 15.0 M and has a density of 0.90 g·mL⁻¹. Calculate the molality of ammonia in this solution.
Ans: 23.3 m

59. Calculate the molality of perchloric acid in 9.2 M HClO₄(aq). The density of this solution is 1.54 g/mL.
Ans: 15 m

60. Calculate the vapor pressure at 25°C of a mixture of benzene and toluene in which the mole fraction of benzene is 0.650. The vapor pressure at 25°C of benzene is 94.6 Torr and that of toluene is 29.1 Torr.
A) 84.4 Torr   B) 124 Torr   C) 51.3 Torr   D) 71.7 Torr   E) 61.5 Torr
Ans: D

61. The vapor pressures of pure carbon disulfide and carbon tetrachloride are 360 and 99.8 Torr, respectively, at 296 K. What is the vapor pressure of a solution containing 50.0 g of each compound?
A) 241 Torr   B) 33.0 Torr   C) 260 Torr   D) 274 Torr   E) 460 Torr
Ans: D

62. The vapor pressures of pure carbon disulfide and carbon tetrachloride are 360 and 99.8 Torr, respectively, at 296 K. What is the vapor pressure of a solution containing 50.0 g of each compound?
A) 260 Torr   B) 274 Torr   C) 241 Torr   D) 33.0 Torr   E) 460 Torr
Ans: B

63. Which of the following 1.0 M solutions contains the most particles?
    A) ethylene glycol
    B) potassium chloride
    C) magnesium sulfate
    D) glucose
    E) sodium sulfate
    Ans: E

64. Which of the following pairs have a van't Hoff i factor of 2?
    A) sodium sulfate and potassium chloride
    B) sodium chloride and magnesium sulfate
    C) magnesium sulfate and ethylene glycol
    D) glucose and sodium chloride
    E) perchloric acid and barium hydroxide
    Ans: B

65. Which of the following pairs have a van't Hoff i factor of 3?
    A) sodium sulfate and potassium chloride
    B) calcium chloride and potassium sulfate
    C) magnesium sulfate and ethylene glycol
    D) glucose and sodium chloride
    E) perchloric acid and barium hydroxide
    Ans: B

66. Which of the following has the lowest freezing point and the highest boiling point?
    A) 2.0 m potassium chloride
    B) 1.5 m magnesium phosphate
    C) 1.0 m sodium chloride
    D) 1.5 m aluminum nitrate
    E) 1.5 m calcium chloride
    Ans: B

67. Which of the following has the lowest freezing point?
    A) 2.0 m sodium perchlorate
    B) 1.0 m potassium phosphate
    C) 1.5 m aluminum nitrate
    D) 1.0 m aluminum sulfate
    E) 1.5 m magnesium phosphate
    Ans: C

68. The freezing point of seawater is about −1.85°C. If seawater is an aqueous solution of sodium chloride, calculate the molality of seawater. The k_f for water is 1.86 K/m.
    A) 1.99 m   B) 0.995 m   C) 3.44 m   D) 3.70 m   E) 0.497 m
    Ans: E

69. The addition of 125 mg of caffeine to 100 g of cyclohexane lowered the freezing point by 0.13 K. Calculate the molar mass of caffeine. The k_f for cyclohexane is 20.1 $K \cdot kg \cdot mol^{-1}$.
    A) 47.8 $g \cdot mol^{-1}$
    B) 481 $g \cdot mol^{-1}$
    C) 96.5 $g \cdot mol^{-1}$
    D) 19.3 $g \cdot mol^{-1}$
    E) 193 $g \cdot mol^{-1}$
    Ans: E

70. Blood, sweat, and tears are about 0.15 M in sodium chloride. Estimate the osmotic pressure of these solutions at 37°C. The gas constant is 0.0821 $L \cdot atm \cdot mol^{-1} \cdot K^{-1}$.
    A) 3.8 atm   B) 11 atm   C) 0.91 atm   D) 1.8 atm   E) 7.6 atm
    Ans: E

71. The osmotic pressure of 1.00 g of a polymer dissolved in benzene to give 200 mL of solution is 1.50 kPa at 25°C. Estimate the average molar mass of the polymer. The gas law constant is 0.0821 $L \cdot atm \cdot mol^{-1} \cdot K^{-1}$.
    A) 41,300 $g \cdot mol^{-1}$
    B) 8260 $g \cdot mol^{-1}$
    C) 693 $g \cdot mol^{-1}$
    D) 1650 $g \cdot mol^{-1}$
    E) 62,000 $g \cdot mol^{-1}$
    Ans: B

72. An animal cell assumes its normal volume when it is placed in a solution with a total solute molarity of 0.3 M. If the cell is placed in a solution with a total solute molarity of 0.1 M,
    A) water enters the cell, causing expansion.
    B) water leaves the cell, causing contraction.
    C) the escaping tendency of water in the cell increases.
    D) no movement of water takes place.
    Ans: A

73. The van't Hoff i of HF is the same as for HCl.
    Ans: False

74. Consider the diagram below.

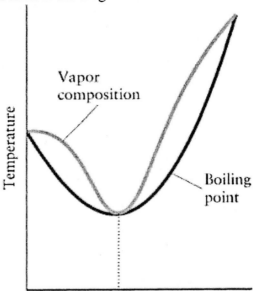

Composition

(a) What is the mixture in the diagram called?

(b) Can the components of this mixture be separated by fractional distillation? Explain.

Ans: (a) minimum-boiling azeotrope

(b) No. In a fractional distillation, the mixture marked by the vertical dotted line boils over first, not the more volatile pure liquid.

75. Solutions in which intermolecular forces are stronger in the solution than in the pure components have negative enthalpies of mixing.

Ans: True

76. Rank the following species in order of increasing vapor pressure at 25°C.

water, diethyl ether, ethanol, mercury

Ans: mercury<water<ethanol<diethyl ether

77. The normal boiling point of ethanol is 78°C. If the vapor pressure of ethanol is 13.3 kPa at 34.9°C, calculate the enthalpy of vaporization of ethanol.

Ans: 42.4 kJ/mol

78. Water and acetone, CH_3COCH_3, both freeze at a higher temperature under pressure. True or False?

Ans: False

79. The critical temperature of N_2 is −147°C. Nitrogen gas cannot be converted to liquid nitrogen at −73°C, even at extremely high pressures. True or False?

Ans: True

80. Consider the phase diagram for CO_2 in the text. A sample of carbon dioxide at 10°C and 10 atm is a liquid. True or False?
Ans: True

81. Which of the following solutes is likely to be most soluble in water?
A) ethanol, CH_3CH_2OH B) carbon tetrachloride, CCl_4 C) Br_2 D) CS_2
Ans: A

82. The solubility of oxygen at a certain temperature is 8 ppm. If the partial pressure of oxygen is doubled at this temperature, the solubility is 16 ppm. True or False?
Ans: True

83. The lattice enthalpy and the enthalpy of hydration are both always positive. True or False?
Ans: False

84. The increase in entropy of the system is responsible for the solubilities of substances that dissolve endothermically. True or False?
Ans: True

85. Because a solute increases the entropy of the liquid phase, while the enthalpy is left unchanged, there is a decrease in the molar free energy of the solvent. True or False?
Ans: True

Chapter 9: Chemical Equilibria

1. For the equilibrium $N_2O_4(g) \rightleftharpoons 2NO_2(g)$, plot, on the same graph, the forward and reverse reaction rates as a function of time. If possible, mark on the graph where equilibrium has been reached.

    Ans: Eventually the rate of the forward reaction equals the rate of the reverse reaction—this is equilibrium.

2. For the equilibrium $CaCO_3(s) \rightleftharpoons CaO(s) + CO_2(g)$, (a) represents the composition at equilibrium at a certain temperature. In (b), a small amount of $Ca^*CO_3(s)$ $Ca^*CO_3(s)$ represents $Ca^{14}CO_3(s)$ or labeled calcium carbonate) has been added. Draw the composition in (c) at equilibrium.

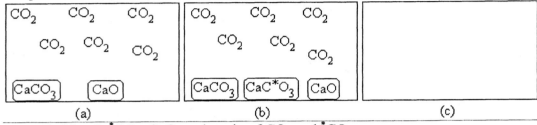

(a)　　　　　　　　(b)　　　　　　　　(c)

    Ans: $CaCO_3$, Ca^*CO_3, CaO, and a mix of CO_2 and $^*CO_2$

3. A 1 L flask containing $D_2(g)$, $N_2(g)$, and $ND_3(g)$ is at equilibrium at 300 K and a 1 L flask of $H_2(g)$, $N_2(g)$, and $NH_3(g)$ is also at equilibrium at 300 K. If the contents of the two flasks are mixed in a 2 L container, write the formulas of all species in the 2 L container at equilibrium.

    Ans: $H_2(g)$, $HD(g)$, $D_2(g)$, $N_2(g)$, $NH_3(g)$, $NHD_2(g)$, $NH_2D(g)$, $ND_3(g)$

4. Pure liquids or solids do not appear in the equilibrium constant expression.

    Ans: True

5. Consider the following reaction at a certain temperature:
    $$Ni(s) + 4CO(g) \rightleftharpoons Ni(CO)_4(g)$$
    Calculate K for this reaction if, at equilibrium, the partial pressures of $CO(g)$ and $Ni(CO)_4(g)$ are 1.25 and 6.65 atm over 1.00 kg of nickel.

    Ans: 2.72

6. Calculate ΔG at 298 K for the reaction
    $$C_2H_5OH(l) \rightarrow C_2H_5OH(g, 0.0263 \text{ bar})$$
    given $\Delta G° = 6.2$ kJ at 298 K.

    A) 6.2 kJ　　B) 15 kJ　　C) 2.8 kJ　　D) −2.8 kJ　　E) −15 kJ

    Ans: D

7. Calculate ΔG at 298 K for the reaction
$$C_2H_5OH(l) \rightarrow C_2H_5OH(g, 0.0400 \text{ bar})$$
given $\Delta G° = 6.2$ kJ at 298 K.
A) 14 kJ   B) 2.7 kJ   C) −14 kJ   D) −1.8 kJ   E) 1.8 kJ
Ans: D

8. Consider the following reaction
$$CuSO_4(s) \rightarrow CuO(s) + SO_3(g)$$
If $\Delta G° = -14.6$ kJ at 950°C for this reaction, calculate ΔG for an $SO_3(g)$ pressure of 20 bar at this temperature.
A) 15.9 kJ   B) 45.1 kJ   C) −14.6 kJ   D) 30.5 kJ   E) −45.1 kJ
Ans: A

9. Consider the following reaction
$$CuSO_4(s) \rightarrow CuO(s) + SO_3(g)$$
If $\Delta G° = -14.6$ kJ at 950°C for this reaction, calculate ΔG for an $SO_3(g)$ pressure of 50 bar at this temperature.
A) 2.68 kJ   B) 25.2 kJ   C) 54.4 kJ   D) 16.3 kJ   E) −54.4 kJ
Ans: B

10. Consider the reaction
$$2CuBr_2(s) \rightarrow 2CuBr(s) + Br_2(g)$$
If the equilibrium vapor pressure of $Br_2(g)$ is 1.43×10^{-5} Torr at 298 K, calculate ΔG at this temperature when $Br_2(g)$ is produced at a pressure of 7.50×10^{-7} Torr.
A) −7.31 kJ   B) 7.31 kJ   C) 39.9 kJ   D) −3.17 kJ   E) −4.15 kJ
Ans: A

11. Consider the reaction
$$2CuBr_2(s) \rightarrow 2CuBr(s) + Br_2(g)$$
If the equilibrium vapor pressure of $Br_2(g)$ is 1.43×10^{-5} Torr at 298 K, calculate ΔG at this temperature when $Br_2(g)$ is produced at a pressure of 7.50×10^{-8} Torr.
A) −5.65 kJ   B) 13.0 kJ   C) 5.65 kJ   D) −3.42 kJ   E) −13.0 kJ
Ans: E

12. If $\Delta G° = 27.1$ kJ at 25°C for the reaction
$$CH_3COOH(aq) + H_2O(l) \rightleftharpoons CH_3COO^-(aq) + H_3O^+(aq)$$
calculate K_a for this reaction at 298 K.
A) 1.15×10^{-11} B) 5.63×10^4 C) 1.78×10^{-5} D) 1.01   E) 9.89×10^{-1}
Ans: C

13. If $\Delta G_r° = 27.1$ kJ·mol^{-1} at 25°C for the dissociation of acetic acid in aqueous solution, calculate K for the reaction below.
$$CH_3COO^-(aq) + H_3O^+(aq) \rightleftharpoons CH_3COOH(aq) + H_2O(l)$$
Ans: 5.63×10^4

14. Consider the following reaction
$$NO(g) + \tfrac{1}{2}O_2(g) \rightarrow NO_2(g)$$
If $\Delta H° = -56.52$ kJ and $\Delta S° = -72.60$ J·K^{-1} at 298 K, calculate the equilibrium constant for the reaction at 298 K.
A) 1.31×10^6 B) 7.63×10^{-7} C) 660   D) 1.22×10^{14} E) 8.08×10^9
Ans: A

15. Consider the following reaction
$$2Fe_2O_3(s) + 3C(s) \rightarrow 4Fe(s) + 3CO_2(g), \ \Delta H° = 462 \text{ kJ}, \ \Delta S° = 558 \text{ J·K}^{-1}$$
Calculate the equilibrium constant for this reaction at 525°C.
A) 3.04×10^{-3} D) 1.9×10^6
B) 8.07×10^{-2} E) 2.18×10^{-2}
C) 5.20×10^{-7}
Ans: B

16. The equilibrium constant for the reaction
$$HNO_2(aq) + H_2O(l) \rightleftharpoons NO_2{-}(aq) + H_3O^+(aq)$$
is 4.3×10^{-4} at 25°C. Will nitrous acid spontaneously dissociate when
$[HNO_2(aq)] = [NO_2{-}(aq)] = [H_3O^+(aq)] = 1.0$ M? Show your calculations.
Ans: no

17. The equilibrium constant for the reaction
$$HNO_2(aq) + H_2O(l) \rightleftharpoons NO_2{-}(aq) + H_3O^+(aq)$$
is 4.3×10^{-4} at 25°C. Will nitrous acid spontaneously dissociate when
$[HNO_2(aq)] = 1.0$ M and $[NO_2{-}(aq)] = [H_3O^+(aq)] = 1.0 \times 10^{-5}$ M?
Show your calculations.
Ans: yes

18. Calculate the equilibrium constant for the reaction below at 25°C.
$$2TiCl_3(s) + 2HCl(g) \rightleftharpoons 2TiCl_4(g) + H_2(g)$$
given $\Delta G° = +46.6$ kJ.
A) 3.8×10^{-98} B) 1.5×10^{-19} C) 6.7×10^{-9} D) 6.6×10^{18} E) 1.5×10^8
Ans: C

19. Calculate the value of K at 700 K for the reaction
$$H_2(g) + I_2(g) \rightleftharpoons 2HI(g)$$
given that $K_c = 54$ at the same temperature.
A) 3100   B) 2.2   C) 54   D) 9.3   E) 1300
Ans: C

20. What is the relation between K and K_c for the reaction below?

    $H_2(g) + I_2(g) \rightleftharpoons 2HI(g)$

    A) $K = K_c$

    B) $K = RTK_c$

    C) $K_c = (RT)^2K$

    D) $K = (RT)^2K_c$

    E) $K_c = RTK$

    Ans: A

21. What is the relationship between K and K_c for the reaction below?

    $2HgO(s) \rightleftharpoons 2Hg(l) + O_2(g)$

    A) $K_c = (RT)^2K$

    B) $K = K_c$

    C) $K_c = RTK$

    D) $K = RTK_c$

    E) $K = (RT)^2K_c$

    Ans: D

22. At 600°C, $K_c = 2.8$ for the reaction

    $2HgO(s) \rightleftharpoons 2Hg(l) + O_2(g)$

    Calculate K at 600°C for this reaction.

    A) 6800   B) 200   C) 1.4×10^4 D) 2.8   E) 138

    Ans: B

23. What is the relationship between K and K_c for the reaction below?

    $NH_4(NH_2CO_2)(s) \rightleftharpoons 2NH_3(g) + CO_2(g)$

    A) $K_c = (RT)^2K$

    B) $K = RTK_c$

    C) $K = (RT)^2K_c$

    D) $K = (RT)^3K_c$

    E) $K_c = (RT)^3K$

    Ans: D

24. At 25°C, $K_c = 1.58 \times 10^{-8}$ for the reaction

    $NH_4(NH_2CO_2)(s) \rightleftharpoons 2NH_3(g) + CO_2(g)$

    Calculate K at 25°C for this reaction.

    A) 3.87×10^{-7}

    B) 2.31×10^{-4}

    C) 9.45×10^{-5}

    D) 5.69×10^{-3}

    E) 1.36×10^{-7}

    Ans: B

25. What is the relationship between K and K_c for the reaction below?

    $N_2(g) + 3H_2(g) \rightleftharpoons 2NH_3(g)$

    A) $K = (RT)^6K_c$

    B) $K_c = (RT)^{-2}K$

    C) $K_c = (RT)^2K$

    D) $K = (RT)^{-2}K_c$

    E) $K = (RT)^2K_c$

    Ans: D

26. At 25°C, $K_c = 4.1 \times 10^8$ for the reaction
$$N_2(g) + 3H_2(g) \rightleftharpoons 2NH_3(g)$$
Calculate K at 25°C for this reaction.
A) 9.7×10^7 B) 6.9×10^5 C) 4.1×10^8 D) 1.7×10^9 E) 2.5×10^{11}
Ans: B

27. At 25°C, $K = 6.9 \times 10^5$ for the reaction
$$N_2(g) + 3H_2(g) \rightleftharpoons 2NH_3(g)$$
Calculate K_c at 25°C for this reaction.
A) 2.8×10^4 B) 6.8×10^5 C) 1.1×10^3 D) 1.7×10^7 E) 4.1×10^8
Ans: E

28. At 700 K, K = 54 for the reaction
$$H_2(g) + I_2(g) \rightleftharpoons 2HI(g)$$
Calculate K_c at 700 K for this reaction.
A) 3.2×10^{-4} B) 0.94   C) 7.7×10^{-4} D) 0.45   E) 54
Ans: E

29. Given: $2SO_2(g) + O_2(g) \rightleftharpoons 2SO_3(g)$
At equilibrium at a certain temperature, the concentrations of $SO_3(g)$, $SO_2(g)$, and $O_2(g)$ are 0.12 M, 0.86 M, and 0.33 M, respectively. Calculate the value of K_c for this reaction.
A) 1.31   B) 0.42   C) 0.014   D) 0.059   E) 0.87
Ans: D

30. Given: $2SO_2(g) + O_2(g) \rightleftharpoons 2SO_3(g)$
At equilibrium at a certain temperature, the concentrations of $SO_3(g)$, $SO_2(g)$, and $O_2(g)$ are 0.24 M, 0.82 M, and 0.33 M, respectively. Calculate the value of K_c for this reaction.
A) 0.89   B) 0.21   C) 0.79   D) 0.26   E) 1.04
Ans: D

31. Given: $C(s) + CO_2(g) \rightleftharpoons 2CO(g)$
At equilibrium at a certain temperature, the partial pressures of $CO(g)$ and $CO_2(g)$ are 1.22 atm and 0.780 atm, respectively. Calculate the value of K for this reaction.
A) 3.13   B) 2.00   C) 1.91   D) 1.56   E) 0.640
Ans: C

32. Given: $C(s) + CO_2(g) \rightleftharpoons 2CO(g)$
At equilibrium at a certain temperature, the partial pressures of $CO(g)$ and $CO_2(g)$ are 1.44 atm and 0.820 atm, respectively. Calculate the value of K for this reaction.
A) 2.53   B) 10.1   C) 1.76   D) 3.08   E) 3.51
Ans: A

33. The equilibrium constant, K, for the reaction
$$2HgO(s) \rightleftharpoons 2Hg(l) + O_2(g)$$
is 1.2×10^{-30}. Calculate K for the reaction
$$1/2 O_2(g) + Hg(l) \rightleftharpoons HgO(s).$$
A) -1.1×10^{-15} D) 9.1×10^{14}
B) 8.3×10^{29} E) 1.1×10^{-15}
C) 4.2×10^{29}
Ans: D

34. Given:
$$SO_2(g) \rightleftharpoons O_2(g) + S(s) \qquad K_c = 2.5 \times 10^{-53}$$
$$SO_3(g) \rightleftharpoons 1/2 O_2(g) + SO_2(g) \qquad K_c = 4.0 \times 10^{-13}$$
Calculate K_c for the reaction
$$2S(s) + 3O_2(g) \rightleftharpoons 2SO_3(g)$$
A) 1.6×10^{103} B) 1.6×10^{80} C) 1.0×10^{130} D) 1.6×10^{40} E) 1.0×10^{65}
Ans: C

35. At 600°C, the equilibrium constant for the reaction
$$2HgO(s) \rightleftharpoons 2Hg(l) + O_2(g)$$
is 2.8. Calculate the equilibrium constant for the reaction
$$1/2 O_2(g) + Hg(l) \rightleftharpoons HgO(s)$$
A) -1.7 B) 1.1 C) 0.60 D) 0.36 E) 1.7
Ans: C

36. Given:
$$4NH_3(g) + 5O_2(g) \rightleftharpoons 4NO(g) + 6H_2O(g) \qquad K$$
Calculate the equilibrium constant for the following reaction.
$$2NO(g) + 3H_2O(g) \rightleftharpoons 2NH_3(g) + 5/2 O_2(g)$$
A) $-0.5K$ B) $-2K$ C) $K^{-1/2}$ D) $-K$ E) K^{-1}
Ans: C

37. Given:
$$4NH_3(g) + 5O_2(g) \rightleftharpoons 4NO(g) + 6H_2O(g) \qquad K$$
Calculate the equilibrium constant for the following reaction.
$$2NH_3(g) + 5/2 O_2(g) \rightleftharpoons 2NO(g) + 3H_2O(g)$$
A) K   B) K^{-1} C) 2K   D) 0.5K   E) $K^{1/2}$
Ans: E

38. Given:
$$P_4(s) + 6Cl_2(g) \rightleftharpoons 4PCl_3(l) \qquad K$$
Calculate the equilibrium constant for the following reaction.
$$2PCl_3(l) \rightleftharpoons 3Cl_2(g) + 1/2 P_4(s)$$
A) $-K^{1/2}$ B) $1/K^{1/2}$ C) $1/K^2$ D) $1/K$ E) $K^{1/2}$
Ans: B

39. The value of Q for a given reaction is a constant.
    Ans: False

40. The reaction free energy $\Delta G_r = RT\ln(Q/K)$. True or False?
    Ans: True

41. At equilibrium, Q = _____.
    Ans: K

42. Which of the following is true?
    A) When the value of Q is large, the equilibrium lies on the product side of the equilibrium reaction.
    B) When the value of K is large, the equilibrium lies on the reactant side of the equilibrium reaction.
    C) A small value of K means that the equilibrium concentrations of the reactants are small compared to the equilibrium concentrations of the products.
    D) A large value of K means that the equilibrium concentrations of products are large compared to the equilibrium concentrations of the reactants.
    E) When the value of K is small, the equilibrium lies on the product side of the equilibrium reaction.
    Ans: D

43. Write the equilibrium constant for
    $2NaBr(aq) + Pb(ClO_4)_2(aq) \rightleftharpoons PbBr_2(s) + 2NaClO_4(aq)$.
    A) $K = [Pb^{2+}][Br^-]^2$
    B) $K = 1/([Pb^{2+}][Br^-]^2)$
    C) $K = [NaClO_4]^2/([NaBr]^2[Pb(ClO_4)_2])$
    D) $K = [PbBr_2]/([Pb^{2+}][Br^-]^2)$
    E) $K = 1/([Pb(ClO_4)_2][NaBr]^2)$
    Ans: B

44. The equilibrium constant, K_c, for the reaction
    $2SO_2(g) + O_2(g) \rightleftharpoons 2SO_3(g)$
    is 11.7 at 1100 K. A mixture of SO_2, O_2, and SO_3, each with a concentration of 0.015 M, was introduced into a container at 1100 K. Which of the following is true?
    A) $SO_2(g)$ and $O_2(g)$ will be formed until equilibrium is reached.
    B) $[SO_3] = 0.045$ M at equilibrium.
    C) $[SO_3] = 0.015$ M at equilibrium.
    D) $SO_3(g)$ will be formed until equilibrium is reached.
    E) $[SO_3] = [SO_2] = [O_2]$ at equilibrium.
    Ans: A

45. The equilibrium constant, K_c, for the reaction

$$2NOCl(g) \rightleftharpoons 2NO(g) + Cl_2(g)$$

is 0.51 at a certain temperature. A mixture of NOCl, NO, and Cl_2 with concentrations 1.3, 1.2, and 0.60 M, respectively, was introduced into a container at this temperature. Which of the following is true?
    A) $Cl_2(g)$ is produced until equilibrium is reached.
    B)   [NOCl] = [NO] = [Cl_2] at equilibrium.
    C)   NOCl(g) is produced until equilibrium is reached.
    D)   [Cl_2] = 0.30 M at equilibrium.
    E)   No apparent reaction takes place.
    Ans: E

46. A mixture consisting of 0.250 M $N_2(g)$ and 0.500 M $H_2(g)$ reaches equilibrium according to the equation below:

$$N_2(g)\ 3H_2(g) \rightleftharpoons 2NH_3(g)$$

At equilibrium, the concentration of ammonia is 0.150 M. Calculate the concentration of $N_2(g)$ at equilibrium.
    A) 0.150 M   B) 0.100 M   C) 0.0750 M   D) 0.0500 M   E) 0.175 M
    Ans: E

47. A mixture consisting of 0.250 M $N_2(g)$ and 0.500 M $H_2(g)$ reaches equilibrium according to the equation below:

$$N_2(g)\ 3H_2(g) \rightleftharpoons 2NH_3(g)$$

At equilibrium, the concentration of ammonia is 0.150 M. Calculate the concentration of $H_2(g)$ at equilibrium.
    A) 0.0750 M   B) 0.350 M   C) 0.425 M   D) 0.275 M   E) 0.150 M
    Ans: D

48. Consider the following reaction:

$$Ni(CO)_4(g) \rightleftharpoons Ni(s) + 4CO(g)$$

If the initial concentration of $Ni(CO)_4(g)$ is 1.0 M, and "x" is the equilibrium concentration of CO(g), what is the correct equilibrium relation?
    A) $K_c = x^4/(1.0 - 4x)$ D) $K_c = x^5/(1.0 - x/4)$
    B) $K_c = x/(1.0 - x/4)$ E) $K_c = 4x/(1.0 - 4x)$
    C) $K_c = x^4/(1.0 - x/4)$
    Ans: C

49. Consider the following reaction:

$$N_2(g)\ 3H_2(g) \rightleftharpoons 2NH_3(g)$$

If the initial concentrations of nitrogen and hydrogen are each 1.0 M, and "x" is the equilibrium concentration of ammonia, what is the correct equilibrium expression?
    Ans: $K_c = (x/2)^2/\{(1.0 - x/2)(1.0 - 3x/2)^3\}$

50. For the following reaction

$$NH_3(g) + H_2S(g) \rightleftharpoons NH_4HS(s)$$

$K_c = 9.7$ at 900 K. If the initial concentrations of $NH_3(g)$ and $H_2S(g)$ are 2.0 M, what is the equilibrium concentration of $NH_3(g)$?

A) 1.9 M   B) 1.7 M   C) 0.20 M   D) 0.10 M   E) 0.32 M

Ans: E

51. For the following reaction

$$NH_3(g) + H_2S(g) \rightleftharpoons NH_4HS(s)$$

$K_c = 9.7$ at 900 K. If the initial concentrations of $NH_3(g)$ and $H_2S(g)$ are 2.0 M, what is the equilibrium concentration of $H_2S(g)$?

A) 1.9 M   B) 0.20 M   C) 1.7 M   D) 0.10 M   E) 0.32 M

Ans: E

52. Consider the following reaction:

$$PCl_5(g) \rightleftharpoons PCl_3(g) + Cl_2(g)$$

At a certain temperature, if the initial concentration of $PCl_5(g)$ is 2.0 M, at equilibrium the concentration of $Cl_2(g)$ is 0.30 M. Calculate the value of K_c at this temperature.

A) 0.064   B) 0.053   C) 0.090   D) 19   E) 0.045

Ans: B

53. Consider the following reaction:

$$PCl_5(g) \rightleftharpoons PCl_3(g) + Cl_2(g)$$

At a certain temperature, if the initial concentration of $PCl_5(g)$ is 3.0 M, at equilibrium the concentration of $Cl_2(g)$ is 0.80 M. Calculate the value of K_c at this temperature.

A) 0.21   B) 0.29   C) 0.64   D) 3.4   E) 0.46

Ans: B

54. For the reaction below

$$2CaSO_4(s) \rightleftharpoons 2CaO(s) + 2SO_2(g) + O_2(g)$$

$K = 0.032$ at 700 K. What is the total pressure starting from pure $CaSO_4(s)$?

A) 0.22 bar   B) 0.011 bar   C) 0.60 bar   D) 0.20 bar   E) 0.40 bar

Ans: C

55. Consider the following reaction at a certain temperature:

$$PCl_5(g) \rightleftharpoons PCl_3(g) + Cl_2(g) \qquad K_c = 0.100$$

At equilibrium, $[PCl_5] = 2.00$ M and $[PCl_3] = [Cl_2] = 1.00$ M. If suddenly 1.00 M $PCl_5(g)$, $PCl_3(g)$, and $Cl_2(g)$ are added, calculate the equilibrium concentration of $Cl_2(g)$.

A) 3.0 M   B) essentially zero   C) 0.65 M   D) 2.75 M   E) 3.35 M

Ans: C

56. Consider the following reaction at a certain temperature:
$$PCl_5(g) \rightleftharpoons PCl_3(g) + Cl_2(g) \qquad K_c = 0.100$$
At equilibrium, $[PCl_5] = 2.00$ M and $[PCl_3] = [Cl_2] = 1.00$ M. If suddenly 1.00 M $PCl_5(g)$, $PCl_3(g)$, and $Cl_2(g)$ are added, calculate the equilibrium concentration of $PCl_5(g)$.
A) 0.65 M   B) 4.35 M   C) 1.35 M   D) essentially zero   E) 2.35 M
Ans: B

57. Consider the following reaction at a certain temperature:
$$PCl_5(g) \rightleftharpoons PCl_3(g) + Cl_2(g) \qquad K_c = 0.100$$
At equilibrium, $[PCl_5] = 2.00$ M and $[PCl_3] = [Cl_2] = 1.00$ M. If suddenly 1.00 M $PCl_3(g)$ and $Cl_2(g)$ are added, calculate the equilibrium concentration of $PCl_5(g)$.
A) essentially 4.00 M   B) 0.58 M   C) 2.58 M   D) 3.42 M   E) 1.42 M
Ans: D

58. Consider the reaction
$$3Fe(s) + 4H_2O(g) \rightleftharpoons 4H_2(g) + Fe_3O_4(s)$$
If the volume of the container is reduced,
A)   the equilibrium constant increases.   D)   more $H_2O(g)$ is produced.
B)   more $H_2(g)$ is produced.   E)   more $Fe(s)$ is produced.
C)   no change occurs.
Ans: C

59. Consider the reaction
$$Na^+(g) + Cl^-(g) \rightleftharpoons NaCl(s)$$
If the temperature is lowered, the (*products or reactants*) are favored.
Ans: products

60. The effect of a volume decrease on the reaction
$$C(s) + H_2O(g) \rightleftharpoons CO(g) + H_2(g)$$
is
A)   that K decreases.   D)   more $H_2O(g)$ produced.
B)   more $CO(g)$ and $H_2(g)$ produced.   E)   that K increases.
C)   no change.
Ans: D

61. Consider the reaction below:
$$F_2(g) \rightleftharpoons 2F(g)$$
(a) Compressing the reaction mixture results in a change in Q. True or False?
(b) Heating the reaction mixture causes the reaction to shift to the left. True or False?
(c) At 1000 K, the equilibrium constant for the reaction is about 10^{-4}. If the reaction is perturbed such that Q = 1, the reaction must shift to the left. True or False?
Ans: (a) True
     (b) False
     (c) False

62. For the reaction
$$N_2O_4(g) \rightleftharpoons 2NO_2(g)$$
Which of the following disturbances will cause an increase in $NO_2(g)$ concentration?
   A)   a decrease in temperature   D)   an increase in pressure
   B)   need ΔH for the reaction to predict   E)   an increase in temperature
   C)   removal of some $N_2O_4(g)$
   Ans: E

63. Consider the reaction below:
$$Ni(s) + 4CO(g) \rightleftharpoons Ni(CO)_4(g)$$
At 30°C and $P_{CO} = 1$ atm, Ni reacts with $CO(g)$ to form $Ni(CO)_4(g)$. At 200°C, $Ni(CO)_4(g)$ decomposes to $Ni(s)$ and $CO(g)$. This means
   A)   adding an inert gas like argon favors the forward reaction.
   B)   the activation energy for the forward reaction is greater than for the reverse reaction.
   C)   the forward reaction is endothermic.
   D)   K at 30°C is greater than K at 200°C.
   E)   a decrease in pressure favors the forward reaction.
   Ans: D

64. The equilibrium constant K for the dissociation of $N_2O_4(g)$ to $NO_2(g)$ is 1700 at 500 K. Predict its value at 300 K. For this reaction, $\Delta H°$ is 56.8 kJ·mol^{-1}.
   A) 1.32×10^{-6} B) 1.11×10^{-4} C) 15.5   D) 0.188   E) 1.54×10^7
   Ans: D

65. Consider the following reaction:
$$2NO_2(g) \rightleftharpoons N_2O_4(g)$$
The equilibrium constant for this reaction will decrease with an increase in temperature. True or False?
   Ans: True

66. For the decomposition of ammonia to nitrogen and hydrogen, the equilibrium constant is 1.47×10^{-6} at 298 K. Calculate the temperature at which K = 1.00. For this reaction, $\Delta H° = 92.38$ kJ·mol^{-1}.
   A) 193 K   B) 353 K   C) 466 K   D) 492 K   E) 219 K
   Ans: C

67. For the decomposition of ammonia to nitrogen and hydrogen, the equilibrium constant is 1.47×10^{-6} at 298 K. Calculate the temperature at which K = 0.0100. For this reaction, $\Delta H° = 92.38$ kJ·mol^{-1}.
   A) 241 K   B) 332 K   C) 59 K   D) 390 K   E) 117 K
   Ans: D

68. Consider the following reaction:
    $$CO(g) + 2H_2(g) \rightleftharpoons CH_3OH(g)$$
    At room temperature, K is approximately 2×10^4, but at a higher temperature K is substantially smaller. Which of the following is true?
    A)   The reaction is endothermic.
    B)   The value of K_c for this reaction is smaller at all temperatures.
    C)   At the higher temperature, more $CH_3OH(g)$ is produced.
    D)   The reaction is exothermic.
    E)   The reaction becomes spontaneous at higher temperatures.
    Ans: D

69. Consider the following reaction:
    $$2HI(g) \rightleftharpoons H_2(g) + I_2(g)$$
    At 298 K, $K_c = 1.3 \times 10^{-3}$, whereas at 783 K, $K_c = 2.2 \times 10^{-2}$. Which of the following is true?
    A)   The reaction is exothermic.
    B) $K = K_c$ at both temperatures.
    C)   At 298 K, $K = 3.2 \times 10^{-2}$.
    D)   At 298 K, the reaction is likely to be spontaneous.
    E)   At 783 K, more $HI(g)$ is produced.
    Ans: B

70. Consider the reaction below:
    $$4NH_3(g) + 7O_2(g) \rightleftharpoons 2N_2O_4(g) + 6H_2O(g)$$
    If, initially, $[NH_3(g)] = [O_2(g)] = 3.60$ M, at equilibrium, $[N_2O_4(g)] = 0.60$ M. Calculate the equilibrium concentrations of all other species.
    Ans: $[NH_3(g)] = 2.40$ M, $[O_2(g)] = 1.50$ M, $[H_2O(g)] = 1.80$ M.

71. Which of the following equilibrium reactions is not affected by changes in pressure?
    A) $2BrCl(g) \rightleftharpoons Br_2(g) + Cl_2(g)$ D) $H_2(g) + I_2(s) \rightleftharpoons 2HI(g)$
    B) $H_2(g) + Br_2(l) \rightleftharpoons 2HBr(g)$ E) $2CO_2(g) \rightleftharpoons 2CO(g) + O_2(g)$
    C) $2H_2O_2(l) \rightleftharpoons 2H_2O(l) + O_2(g)$
    Ans: A

72. Consider the reaction below:
    $$4NH_3(g) + 3O_2(g) \rightleftharpoons 2N_2(g) + 6H_2O(g), K = 10^{80} \text{ at a certain temperature.}$$
    Initially, all reactants and products have concentrations equal to 12 M. At equilibrium, the approximate concentration of oxygen is
    A) 6 M.   B) 0 M.   C) 3 M.   D) 12 M.   E) 18 M.
    Ans: C

73. Consider the reaction below:

$4NH_3(g) + 3O_2(g) \rightleftharpoons 2N_2(g) + 6H_2O(g)$, K = 10^{80} at a certain temperature.
Initially, all reactants and products have concentrations equal to 12 M. At equilibrium, the approximate concentration of ammonia is
A) 6 M.   B) 3 M.   C) 12 M.   D) 18 M.   E) 0 M.
Ans: E

74. For the reaction $2NOCl(g) \rightleftharpoons 2NO(g) + Cl_2(g)$, if, initially, $[NOCl(g)]$ = 2.8 M, at equilibrium $[NO(g)]$ = 1.2 M. Calculate the equilibrium concentration of $NOCl(g)$.
Ans: 1.6 M

75. For the reaction $2NOCl(g) \rightleftharpoons 2NO(g) + Cl_2(g)$, K = 98 at a certain temperature. If the equilibrium concentrations in a 1 L container are $[NOCl(g)]$ = 1.0 M, $[NO(g)]$ = 3.5 M and $[Cl_2(g)]$ = 8.0 M, and 2.0 moles of each gas is added, in which direction does the reaction shift?
Ans: to the left

76. If the equilibrium constant for the reaction $Ni(s) + 4CO(g) \rightleftharpoons Ni(CO)_4(g)$ is 2.72 at a certain temperature, what is the equilibrium constant for the following reaction at the same temperature?
$Ni(CO)_4(g) \rightleftharpoons Ni(s) + 4CO(g)$
Ans: 0.368

77. If a reaction mixture that is not at equilibrium contains more products than reactants, ΔG > 0 for the forward reaction. True or False?
Ans: True

78. From a plot of Gibbs Free Energy versus Progress of Reaction, the sign of ΔG_r at any point along the curve is given by the slope of the curve. True or False?
Ans: True

79. For any reaction at equilibrium, ΔG < 0. True or False?
Ans: False

80. For a pure solid or liquid, the molar free energy always has its standard value. True or False?
Ans: True

81. The vapor pressure of acetic acid at 25°C is 16 Torr. ΔG_r for the reaction
$CH_3COOH(l) \rightleftharpoons CH_3COOH(g)$
at 25°C is
A) 0   B) +9.57 kJ·mol^{-1} C) −9.57 kJ·mol^{-1} D) +1.85 kJ·mol^{-1}
Ans: A

82. An endothermic reaction is most likely to have a small equilibrium constant if ΔS_r° is small. True or False?
   Ans: True

83. At 25°C, ΔG_r° for the reaction $2SO_2(g) + O_2(g) \rightleftharpoons 2SO_3(g)$ is -141.74 kJ·mol^{-1}.
   Calculate the value of K_c for this reaction.
   A) 1.74×10^{26} B) 7.01×10^{24} C) 2.65×10^{12} D) 6.56×10^{13}
   Ans: A

84. For the reaction $Br_2(g) \rightleftharpoons 2Br(g)$, $\Delta G_r^\circ = +161.69$ kJ·mol^{-1} at 25°C. Calculate the value of K_c for this reaction.
   A) 0.0378 B) 1.12×10^{-27} C) 1.83×10^{-30} D) 4.54×10^{-29}
   Ans: C

85. Consider the following reaction:
   $2SO_2(g) + O_2(g) \leftrightarrow 2SO_3(g)$
   At equilibrium at a certain temperature, the partial pressures of $SO_2(g)$, $O_2(g)$, and $SO_3(g)$ are 0.0012, 0.18, and 2.2 bar. Argon is introduced into the reaction vessel until the partial pressure of argon is 5.0 bar. Predict the final partial pressures of the reactants and products.
   Ans: They all stay the same.

Chapter 10: Acids and Bases

1. All of the following are strong bases in water except
   A) $NaHCO_3$ B) CaO   C) Na_2O D) Na_2SO_4
   Ans: A

2. What is the conjugate acid of O^{2-}?
   Ans: OH^-

3. The conjugate acid of HPO_4^{2-} is
   A) HPO_4^{2-}.   B) PO_4^{3-}.   C) $H_2PO_4^-$.   D) H_3O^+.   E) H_3PO_4.
   Ans: C

4. The conjugate base of $H_2PO_4^-$ is
   A) PO_4^{3-}.   B) OH^-.   C) H_3PO_4.   D) $H_2PO_4^-$.   E) HPO_4^{2-}.
   Ans: E

5. The conjugate base of ammonia is
   A) NH_2OH.   B) NH_2^-.   C) NH_4^+.   D) NH_3.   E) OH^-.
   Ans: B

6. The conjugate base of OH^- is
   A) H^+.   B) OH^-.   C) O^{2-}.   D) H_3O^+.   E) H_2O.
   Ans: C

7. In the following reaction
   $$SO_2(g) + H_2O(l) \rightarrow H_2SO_3(aq),$$
   identify the Lewis acid and base.
   Ans: Lewis acid, SO_2; Lewis base, H_2O.

8. When sulfur trioxide dissolves in water, sulfuric acid is produced. An intermediate in the reaction is H_2O-SO_3. In the reaction of the intermediate to produce sulfuric acid,
   A)   water acts both as an acid and a base.
   B)   water acts as a proton donor only.
   C)   water acts as a proton acceptor only.
   D)   the intermediate undergoes an intramolecular rearrangement to form the product.
   Ans: A

9. All of the following have amphoteric oxides except
   A) PbO.   B) MgO.   C) SnO.   D) Al_2O_3.   E) BeO.
   Ans: B

10. Write the autoprotolysis reaction for liquid ammonia.
    Ans: $2NH_3(l) \rightleftharpoons NH_2^-(\text{solvated}) + NH_4^+(\text{solvated})$

11. Calculate the hydrogen ion concentration for an aqueous solution that has a pH of 3.45.
    A) 0.54 M   B) 3.5×10^{-4} M   C) 2.8×10^{-11} D) 3.2×10^{-2} M   E) 1.22 M
    Ans: B

12. Calculate the hydroxide ion concentration for an aqueous solution that has a pH of 3.45.
    A) 3.2×10^{-2} M   D) 2.8×10^{-11} M
    B) 0.54 M   E) 2.6×10^{-5} M
    C) 3.5×10^{-4} M
    Ans: D

13. Which of the following is the strongest acid?
    A) HCN ($pK_a = 9.31$)   D) CH_3COOH ($pK_a = 4.75$)
    B) HIO_3 ($pK_a = 0.77$)   E) HNO_2 ($pK_a = 3.37$)
    C) HF ($pK_a = 3.45$)
    Ans: B

14. Which of the following is the strongest base?
    A) methylamine ($pK_b = 3.44$)   D) ammonia ($pK_b = 4.75$)
    B) morphine ($pK_b = 5.79$)   E) pyridine ($pK_b = 8.75$)
    C) urea ($pK_b = 13.90$)
    Ans: A

15. Which of the following is the weakest acid?
    A) HNO_3 B) HBr   C) HCl   D) HF   E) HI
    Ans: D

16. Which of the following produces the strongest conjugate base?
    A) HF ($pK_a = 3.45$)   D) CH_3COOH ($pK_a = 4.75$)
    B) HClO ($pK_a = 7.53$)   E) HIO ($pK_a = 10.64$)
    C) HCOOH ($pK_a = 3.75$)
    Ans: E

17. What is the pK_a of the conjugate acid of hydrazine, given that the pK_b of hydrazine is 5.77? Write out the formula of the conjugate acid of hydrazine.
    Ans: $pK_a = 8.23$ for $N_2H_3^+$

18. Strong acids are leveled in water to the strength of the acid H_3O^+.
    Ans: True

19. The pH of a 0.0050 M aqueous solution of calcium hydroxide is
    A) 11.40   B) 2.00   C) 12.00   D) 12.70   E) 11.70
    Ans: C

20. In a solution labeled "0.10 M HNO_3," which of the following is correct?
    A) $[HNO_3] = 0.10$ M
    B) $[H_3O^+] = 0.10$ M, $[NO_3^-] = 0.10$ M
    C) $[H_3O^+] = 0.090$ M, $[NO_3^-] = 0.010$ M
    D) $[HNO_3] = 0.050$ M, $[H_3O^+] = 0.050$ M, $[NO_3^-] = 0.050$ M
    E) $[H_3O^+] = 0.10$ M, $[OH^-] = 1.0 \times 10^{-7}$ M
    Ans: B

21. If the value of K_b for pyridine is 1.8×10^{-9}, calculate the equilibrium constant for
    $$C_5H_5NH^+(aq) + H_2O(l) \rightleftharpoons C_5H_5N(aq) + H_3O^+(aq)$$
    A) -1.8×10^{-9} B) 1.8×10^{-16} C) 5.6×10^8 D) 1.8×10^{-9} E) 5.6×10^{-6}
    Ans: E

22. Which of the following 0.10 M aqueous solutions gives the lowest pH?
    A) CCl_3COOH ($pK_a = 0.52$)
    B)   Because all are acids, the pH is the same for all solutions.
    C)   HF ($pK_a = 3.45$)
    D) CH_3COOH ($pK_a = 4.75$)
    E) H_3PO_4 ($pK_{a1} = 2.12$)
    Ans: A

23. In liquid ammonia, the base B is a strong base if it is a stronger proton acceptor than NH_2^-.
    Ans: True

24. In liquid ammonia, the acid HB is a strong acid if it is a weaker proton donor than NH_4^+.
    Ans: False

25. Which of the following is the strongest acid?
    A) CH_3CH_2OH D) $CH_2ClCOOH$
    B) CH_3COOH E) CCl_3COOH
    C) $CHCl_2COOH$
    Ans: E

26. Which of the following has the highest pK_a?
    A) HClO   B) $HClO_3$ C) HBrO   D) HIO   E) $HClO_4$
    Ans: D

27. Which of the following aqueous solutions gives a pH greater than 7?
    A) 10^{-8} M NH_4Cl
    B)   None of the solutions gives a pH greater than 7.
    C) 10^{-8} M CH_3COOH
    D) 10^{-8} M HCl
    E) 10^{-8} M HCOOH
    Ans: B

28. The pH of 0.80 M benzenesulfonic acid is 0.51. What is the percentage ionization of benzenesulfonic acid?
    A) 25%   B) 39%   C) 51%   D) 5.0%   E) 64%
    Ans: B

29. The pH of 0.010 M aniline(aq) is 8.32. What is the percentage aniline protonated?
    A) 2.1%   B) 0.69%   C) 0.021%   D) 0.12%   E) 0.21%
    Ans: C

30. The pH of 0.800 M aqueous benzenesulfonic acid is 0.51. What is the value of K_a for benzenesulfonic acid?
    A) 0.19   B) 0.12   C) 0.90   D) 0.44   E) 0.51
    Ans: A

31. The pH of 0.10 M pyridine(aq) is 9.13. What is the value of K_b for pyridine?
    A) 2.7×10^{-4} B) 7.4×10^{-10} C) 2.7×10^{-5} D) 1.8×10^{-10} E) 1.8×10^{-9}
    Ans: E

32. What is the pH of 0.25 M HBrO(aq) ($pK_a = 8.69$)?
    A) 0.60   B) 5.90   C) 8.10   D) 9.30   E) 4.65
    Ans: E

33. What is the pH of 0.025 M $(CH_3)_3N$(aq) ($K_b = 6.5 \times 10^{-5}$)?
    A) 11.91   B) 12.40   C) 11.11   D) 8.29   E) 9.81
    Ans: C

34. For a 0.10 M solution of a weak acid, HA, with $pK_a = 10$, which of the following is true?
    A) $[HA] \cong 0$ D) $[HA] = K_a$
    B) $[HA] = [A^-]$ E) $[HA] \neq [H_3O^+]$
    C) $[HA] = [H_3O^+]$
    Ans: E

35. If the pK_a of acetic acid is 4.75, the pK_a of CH_3CH_2OH is
    A)   about 4.   D)   also 4.75.
    B)   much less than 4.75.   E)   about 7.
    C)   about 16.
    Ans: C

36. When CaO(s) is dissolved in water, which of the following is true?
    A) The solution contains O^{2-}(aq), OH^-(aq), and Ca^{2+}(aq).
    B) The solution contains CaO(aq).
    C) CaO(s) does not dissolve in water.
    D) The solution contains O^{2-}(aq) and Ca^{2+}(aq).
    E) The solution contains OH^-(aq) and Ca^{2+}(aq).
    Ans: E

37. Which of the following 0.10 M aqueous solutions has the lowest pH?
    A) $B(OH)_3$ B) HIO   C) $C_2H_5NH_3Cl$ D) C_6H_5OH
    Ans: A

38. Write the charge balance equation for a dilute aqueous solution of HI.
    A) $[I^-] = [OH^-] + [H_3O^+]$
    B) $[H_3O^+] = [OH^-]$
    C) $[H_3O^+] = [I^-]$
    D) $[H_3O^+] = [I^-] + [OH^-]$
    E) $[HI]_{initial} = [I^-]$
    Ans: D

39. Estimate the pH of 10^{-7} M $HClO_4$(aq).
    A) 6.8   B) 8.0   C) 1.0   D) 5.0   E) 7.0
    Ans: A

40. Write the charge balance equation for a dilute aqueous solution of KOH.
    A) $[KOH]_{initial} = [K^+]$
    B) $[OH^-] = [H_3O^+] + [K^+]$
    C) $[H_3O^+] = [OH^-]$
    D) $[K^+] = [OH^-] + [H_3O^+]$
    E) $[OH^-] = [K^+]$
    Ans: B

41. Estimate the pH of 10^{-7} M KOH(aq).
    A) 6.9   B) 9   C) 13   D) 7.2   E) 7.0
    Ans: D

42. The K_a of phenol is 1.3×10^{-10}. For a solution labeled "1.0×10^{-3} M aqueous phenol,"
    A) $[H_3O^+] = [H_3O^+]^2/[phenol]_{initial}$.
    B) $[H_3O^+] \ll [phenol]_{initial}$.
    C) $[H_3O^+] > 10^{-6}$.
    D) pH ~ 4.
    E) $K_w/[H_3O^+] \gg [phenol]_{initial}$.
    Ans: B

43. The K_a of phenol is 1.3×10^{-10}. For a solution labeled "1.0×10^{-3} M aqueous phenol,"
    A) $K_w/[H_3O^+] \gg [phenol]_{initial}$.
    B) $K_w/[H_3O^+] \ll [phenol]_{initial}$.
    C) $[H_3O^+] > 10^{-6}$.
    D) pH ~ 4.
    E) $[H_3O^+] = [H_3O^+]^2/[phenol]_{initial}$.
    Ans: B

44. Write the charge balance equation for a solution that is 0.0010 M phenol(aq). Let phenol be represented by HA(aq).

    A) $[H_3O^+] = [OH^-]$

    B) $K_w = [H_3O^+][OH^-]$

    C) $[H_3O^+] = [OH^-] + [A^-]$

    D) $K_a = K_w/K_b$

    E) $0.0010 = [HA] + [A^-]$

    Ans: C

45. A 0.0010 M solution of a weak acid, HA, with $K_a = 2 \times 10^{-10}$ produces $[H_3O^+] < 10^{-6}$ M. Which of the following equations can be used to determine $[H_3O^+]$?

    A) The acid is so weak that the pH is about 7.

    B) $[H_3O^+]^2 + K_a[H_3O^+] - [HA]_{initial}K_a = 0$

    C) $[H_3O^+] = (K_w + K_a[HA]_{initial})^{\frac{1}{2}}$

    D) $[H_3O^+] = [HA]_{initial}$

    E) $[H_3O^+] = (K_a[HA]_{initial})^{\frac{1}{2}}$

    Ans: C

46. The following 0.1 M aqueous solutions are arranged in order of increasing pH, with the highest pH on the far right.

HNO_3	HCOOH		KBr	KNO_2

    Which one of the following 0.10 M aqueous solutions should be placed in the empty box?

    A) $NaHSO_4$ B) KF C) HNO_2 D) CH_3NH_2 E) $(CH_3)_3NHCl$

    Ans: E

47. The following 0.1 M aqueous solutions are arranged in order of increasing pH, with the highest pH on the far right.

$HClO_4$		KNO_3	KNO_2	RbOH

    Which one of the following 0.10 M aqueous solutions should be placed in the empty box?

    A) $CuSO_4$ B) $NaNO_2$ C) CH_3NH_2 D) $NaHCO_3$ E) Na_2HPO_4

    Ans: A

48. The following 0.1 M aqueous solutions are arranged in order of increasing pH, with the highest pH on the far right.

C_5H_5NHCl	HClO	KNO_3	NaClO	

    Which one of the following 0.10 M aqueous solutions should be placed in the empty box?

    A) NaCN B) CH_3COOH C) KNO_2 D) NaBr E) $NaHSO_4$

    Ans: A

49. The following 0.1 M aqueous solutions are arranged in order of increasing pH, with the highest pH on the far right.

HNO$_3$	C$_5$H$_5$NHCl	HClO		NaClO

Which one of the following 0.10 M aqueous solutions should be placed in the empty box?
A) NH$_4$Cl   B) NaCN   C) Al$_2$(SO$_4$)$_3$ D) CH$_3$NH$_2$ E) CH$_3$COOH
Ans: A

50. The following 0.1 M aqueous solutions are arranged in order of increasing pH, with the highest pH on the far right.

HClO$_4$		(CH$_3$)$_3$NHCl	KBr	KNO$_2$

Which one of the following 0.10 M aqueous solutions should be placed in the empty box?
A) NaI   B) HCOOH   C) C$_6$H$_5$NH$_2$ D) CH$_3$NH$_3$Cl   E) NaClO
Ans: B

51. The equation that represents K$_{a2}$ for sulfurous acid is
A)   HSO$_3^-$(aq) + H$_2$O(l) $\rightleftharpoons$ H$_2$SO$_3$(aq) + OH$^-$(aq).
B)   HSO$_3^-$(aq) + H$_2$O(l) $\rightleftharpoons$ SO$_3^{2-}$(aq) + H$_3$O$^+$(aq).
C)   H$_2$SO$_3$(aq) + 2H$_2$O(l) $\rightleftharpoons$ SO$_3^{2-}$(aq) + 2H$_3$O$^+$(aq).
D)   SO$_3^{2-}$(aq) + H$_2$O(l) $\rightleftharpoons$ HSO$_3^-$(aq) + OH$^-$(aq).
E)   H$_2$SO$_3$(aq) + H$_2$O(l) $\rightleftharpoons$ HSO$_3^-$(aq) + H$_3$O$^+$(aq).
Ans: B

52. The equation that represents K$_{a2}$ for phosphoric acid is
A)   HPO$_4^{2-}$(aq) + H$_2$O(l) $\rightleftharpoons$ PO$_4^{3-}$(aq) + H$_3$O$^+$(aq)
B)   H$_2$PO$_4^-$(aq) + H$_2$O(l) $\rightleftharpoons$ HPO$_4^{2-}$(aq) + H$_3$O$^+$(aq).
C)   H$_3$PO$_4$(aq) + 2H$_2$O(l) $\rightleftharpoons$ HPO$_4^{2-}$(aq) + 2H$_3$O$^+$(aq).
D)   HPO$_4^{2-}$(aq) + H$_2$O(l) $\rightleftharpoons$ H$_2$PO$_4^-$(aq) + OH$^-$(aq).
E)   H$_3$PO$_4$(aq) + H$_2$O(l) $\rightleftharpoons$ H$_2$PO$_4^-$(aq) + H$_3$O$^+$(aq).
Ans: B

53. For a solution labeled "0.10 M H$_2$SO$_4$(aq),"
A)   [HSO$_4^-$] is greater than 0.10 M.   D)   the pH equals 1.0.
B)   the pH is less than 1.0.   E)   the pH is greater than 1.0.
C)   [SO$_4^{2-}$] = 0.10 M.
Ans: B

54. For a solution labeled "0.10 M H$_3$PO$_4$(aq),"
A)   [H$_2$PO$_4^-$] is greater than 0.10 M.   D)   [H$^+$] = 0.10 M.
B)   [H$^+$] = 0.30 M.   E)   [H$^+$] is less than 0.10 M.
C)   [PO$_4^{3-}$] = 0.10 M.
Ans: E

55. For a solution labeled "0.10 M H_2SO_3(aq)," pK_{a1} = 1.81 and pK_{a2} = 6.91, which of the following is true?
    A) $[H^+]$ = 0.2 M.   D) The pH is 0.70.
    B) The pH is 1.0.   E) The pH is about 4.4.
    C) The pH is about 1.5.
    Ans: C

56. Calculate $[H^+]$ for a solution labeled "0.0500 M H_2SO_3(aq)" (pK_{a1} = 1.81, pK_{a2} = 6.91).
    A) 0.021 M   B) 0.029 M   C) 0.015 M   D) 0.025 M   E) 0.050 M
    Ans: A

57. Calculate the equilibrium concentration of sulfurous acid in a solution labeled "0.100 M H_2SO_3(aq)" (pK_{a1} = 1.81, pK_{a2} = 6.91).
    A) 0.068 M   B) 0.015 M   C) 0.100 M   D) 0.050 M   E) 0.032 M
    Ans: A

58. If pK_{a1} and pK_{a2} for H_2CO_3 are 6.37 and 10.25, respectively, calculate the equilibrium constant for the reaction below:
    $$H_2CO_3(aq) + 2H_2O(l) \rightleftharpoons 2H_3O^+(aq) + CO_3{}^{2-}(aq)$$
    A) 4.1×10^{-11} D) 2.3×10^{-8}
    B) 4.3×10^{-7} E) 2.4×10^{-17}
    C) 5.6×10^{-11}
    Ans: E

59. If pK_{a1} and pK_{a2} for H_2S are 6.88 and 14.15, respectively, calculate the equilibrium constant for the reaction below:
    $$H_2S(aq) + 2H_2O(l) \rightleftharpoons 2H_3O^+(aq) + S^{2-}(aq)$$
    A) 1.3×10^{-7} B) 1.1×10^{-7} C) 7.7×10^{-8} D) 9.2×10^{-22} E) 7.1×10^{-15}
    Ans: D

60. Calculate the equilibrium constant for the reaction
    $$HS^-(aq) + H_2O(l) \rightleftharpoons H_2S(aq) + OH^-(aq),$$
    given K_{a1} = 1.3×10^{-7} and K_{a2} = 7.1×10^{-15} for H_2S.
    A) 1.3×10^{-7} B) 7.7×10^{-8} C) 9.2×10^{-22} D) 7.1×10^{-15} E) 1.4
    Ans: B

61. Calculate the equilibrium constant for the reaction
    $$S^{2-}(aq) + H_2O(l) \rightleftharpoons HS^-(aq) + OH^-(aq),$$
    given K_{a1} = 1.3×10^{-7} and K_{a2} = 7.1×10^{-15} for H_2S.
    A) 1.3×10^{-7} B) 9.2×10^{-22} C) 7.7×10^{-8} D) 7.1×10^{-15} E) 1.4
    Ans: E

62. Estimate the pH of 0.10 M $Na_2HPO_4(aq)$ given $pK_{a1} = 2.12$, $pK_{a2} = 7.21$, and $pK_{a3} = 12.68$ for phosphoric acid.
    A) 12.68   B) 9.94   C) 7.40   D) 4.67   E) 2.12
    Ans: B

63. The pH of 0.10 and 0.40 M $NaHCO_3(aq)$ is 8.31 for both solutions. True or False?
    Ans: True

64. The pH of 0.010 M $H_3PO_4(aq)$ is 2.24. Estimate the concentration of HPO_4^{2-} in the solution. For H_3PO_4, the values of K_{a1}, K_{a2}, and K_{a3} are 7.6×10^{-3}, 6.2×10^{-8}, and 2.1×10^{-13}, respectively.
    A) 5.8×10^{-3} M   D) 6.2×10^{-8} M
    B) 7.6×10^{-3} M   E) 2.1×10^{-13} M
    C)   0.010 M
    Ans: D

65. The pH of 0.010 M $H_3PO_4(aq)$ is 2.24. Estimate the concentration of PO_4^{3-} in the solution. For H_3PO_4, the values of K_{a1}, K_{a2}, and K_{a3} are 7.6×10^{-3}, 6.2×10^{-8}, and 2.1×10^{-13}, respectively.
    A) 5.8×10^{-3} M   D) 6.2×10^{-8} M
    B) 2.1×10^{-13} M   E) 2.3×10^{-18} M
    C) 7.6×10^{-3} M
    Ans: E

66. The amino acid alanine, $HOOC-CH(CH_3)NH_3+$, has $K_{a1} = 4.5 \times 10^{-3}$ and $K_{a2} = 1.4 \times 10^{-10}$. Calculate $\alpha(^-OOC-CH(CH_3)NH_3+)$ at pH 10.
    A) 0.42   B) 0.29   C) 1.0   D) 0   E) 0.58
    Ans: A

67. The amino acid alanine, $HOOC-CH(CH_3)NH_3+$, has $K_{a1} = 4.5 \times 10^{-3}$ and $K_{a2} = 1.4 \times 10^{-10}$. Calculate $\alpha(^-OOC-CH(CH_3)NH_3+)$ at pH 3.
    A) 0   B) 0.82   C) 0.18   D) 0.29   E) 0.58
    Ans: B

68. The amino acid alanine, $HOOC-CH(CH_3)NH_3+$, has $K_{a1} = 4.5 \times 10^{-3}$ and $K_{a2} = 1.4 \times 10^{-10}$. Calculate $\alpha(HOOC-CH(CH_3)NH_3+)$ at pH 3.
    A) 0   B) 0.82   C) 0.58   D) 0.29   E) 0.18
    Ans: E

69. The amino acid methionine, $HOOC-CH(CH_2CH_2SCH_3)NH_3^+$, has $pK_{a1} = 2.2$ and $pK_{a2} = 9.1$. If this amino acid is represented by H_2L^+, the major species at pH 6 is
    A) HL   B) H_2L^+ C) L^- D) HL and L^-
    Ans: A

70. For a solution of phosphoric acid, write the equation for $\alpha(HPO_4^{2-})$.
    A) $K_{a3}/([H_3O^+]^3 + K_{a1}[H_3O^+]^2 + K_{a1}K_{a2}[H_3O^+] + K_{a1}K_{a2}K_{a3})$
    B) $[H_3O^+]/([H_3O^+]^3 + K_{a1}[H_3O^+]^2 + K_{a1}K_{a2}[H_3O^+] + K_{a1}K_{a2}K_{a3})$
    C) $K_{a1}K_{a2}[H_3O^+]/([H_3O^+]^3 + K_{a1}[H_3O^+]^2 + K_{a1}K_{a2}[H_3O^+] + K_{a1}K_{a2}K_{a3})$
    D) $K_{a1}K_{a2}K_{a3}/([H_3O^+]^3 + K_{a1}[H_3O^+]^2 + K_{a1}K_{a2}[H_3O^+] + K_{a1}K_{a2}K_{a3})$
    E) $K_{a1}[H_3O^+]^2/([H_3O^+]^3 + K_{a1}[H_3O^+]^2 + K_{a1}K_{a2}[H_3O^+] + K_{a1}K_{a2}K_{a3})$
    Ans: C

71. The fractional composition diagram for the amino acid alanine is given below.

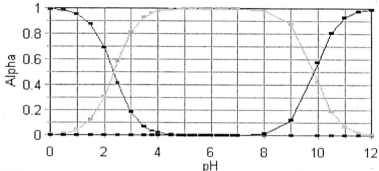

    Write the structure of the dominant species at pH 1, 6, and 12, respectively.
    Ans: $HOOC\text{-}CH(CH_3)NH_3^+$, $^-OOC\text{-}CH(CH_3)NH_3^+$ and $^-OOC\text{-}CH(CH_3)NH_2$.

72. The fractional composition diagram for the amino acid alanine is given below.

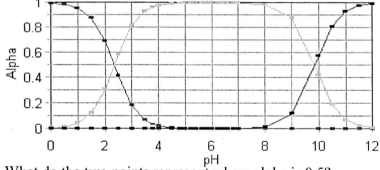

    What do the two points represent where alpha is 0.5?
    Ans: pK_{a1} and pK_{a2}

73. If $\alpha(HSO_3^-) = 0.83$ at pH 2.5, what are $\alpha(H_2SO_3)$ and $\alpha(SO_3^{2-})$ at this pH? For H_2SO_3, pK_{a1} and pK_{a2} are 1.81 and 6.91, respectively.
    A) $\alpha(H_2SO_3) \sim 0$ and $\alpha(SO_3^{2-}) = 0.17$
    B) $\alpha(H_2SO_3) = 0.415$ and $\alpha(SO_3^{2-}) \sim 0$
    C) $\alpha(H_2SO_3) = 0.0.085$ and $\alpha(SO_3^{2-}) = 0.085$
    D) $\alpha(H_2SO_3) = 0.17$ and $\alpha(SO_3^{2-}) \sim 0$
    E) $\alpha(H_2SO_3) = 0.17$ and $\alpha(SO_3^{2-}) \sim 1$
    Ans: D

74. If $\alpha(HSO_3^-) = 0.45$ at pH 7.0, what are $\alpha(H_2SO_3)$ and $\alpha(SO_3^{2-})$ at this pH? For H_2SO_3, pK_{a1} and pK_{a2} are 1.81 and 6.91, respectively.
    A) $\alpha(H_2SO_3) = 0.225$ and $\alpha(SO_3^{2-}) = 0.55$
    B) $\alpha(H_2SO_3) = 0.55$ and $\alpha(SO_3^{2-}) \sim 0$
    C) $\alpha(H_2SO_3) \sim 0$ and $\alpha(SO_3^{2-}) = 0.225$
    D) $\alpha(H_2SO_3) \sim 0$ and $\alpha(SO_3^{2-}) = 0.55$
    E) $\alpha(H_2SO_3) = 0.45$ and $\alpha(SO_3^{2-}) = 0.55$
    Ans: D

75. The boxes below contain a series of 0.1 M aqueous solutions of increasing pH where A is the solution of lowest pH and E is the solution of highest pH.

A	B	C	D	E

    Match each box with the correct compound.
        phenol, $pK_a = 9.89$
        cyanide ion, $pK_b = 4.69$
        pyridine, $pK_b = 8.75$
        hydrogen sulfate ion, $pK_a = 1.92$
        sodium nitrate
    Ans: phenol, $pK_a = 9.89$ (B)
        cyanide ion, $pK_b = 4.69$ (E)
        pyridine, $pK_b = 8.75$ (D)
        hydrogen sulfate ion $pK_a = 1.92$ (A)
        sodium nitrate (C)

76. All of the following are Lewis bases except
    A) OH^- B) H_2O C) SO_3 D) Br^-
    Ans: C

77. Both H_2O and OH^- can act as a Brønsted acid and a Brønsted base but not as a Lewis acid. True or False?
    Ans: True

78. All of the following are Lewis acids except
    A) SO_3 B) H_3O^+ C) NO_2^- D) BF_3
    Ans: B

79. What is the molarity of OH^- in a solution labeled "0.0018 M barium hydroxide?"
    A) 0.0018 M   B) 0.0036 M   C) 0.00090 M   D) 0.0072 M
    Ans: B

80. All of the following acids have the same strength in water except
    A) HNO_3 B) $HClO_3$ C) HBr D) HF
    Ans: D

81. Bond polarity tends to dominate the trend of acid strengths for binary acids of elements of the same period. True or False?
    Ans: True

82. Which of the following is the weakest acid?
    A) HNO_3 B) $HClO_4$ C) $HClO_2$ D) $HClO$
    Ans: D

83. All of the following 0.1 M aqueous solutions are acidic except
    A) Na_2SO_4 B) $Cr(ClO_4)_3$ C) C_6H_5OH D) $C_2H_5NH_3Cl$
    Ans: A

84. Is a 0.1 M aqueous solution of HPO_4^{2-} acidic, basic, or neutral? Prove your answer using the appropriate equilibrium constants.
    Ans: basic, $K_b\,(HPO_4^{2-}) = 1.6 \times 10^{-7}$ and $K_a\,(HPO_4^{2-}) = 2.1 \times 10^{-13}$

85. In a solution that is labeled "0.10 M H_3PO_4(aq)," $[H_3O^+] = 0.024$ M. Match the species below with their concentrations.

    H_3PO_4 6.2×10^{-8}
    $H_2PO_4^-$ 8.0×10^{-2}
    HPO_4^{2-} 5.4×10^{-19}
    PO_4^{3-} 2.4×10^{-2}

    Ans: $[H_3PO_4] = 8.0 \times 10^{-2}$, $[H_2PO_4^-] = 2.4 \times 10^{-2}$, $[HPO_4^{2-}] = 6.2 \times 10^{-8}$, $[PO_4^{3-}] = 5.4 \times 10^{-19}$

Chapter 11: Aqueous Equilibria

1. When ammonium chloride is added to NH_3(aq),
    A)   the pH of the solution does not change.
    B)   the pH of the solution increases.
    C)   the pH of the solution decreases.
    D)   the K_b increases.
    E)   the equilibrium concentration of NH_3(aq) decreases.
    Ans: C

2. When sodium formate is added to HCOOH(aq),
    A)   the equilibrium concentration of HCOOH(aq) decreases.
    B)   the pH of the solution increases.
    C)   the K_a increases.
    D)   the pH of the solution does not change.
    E)   the pH of the solution decreases.
    Ans: B

3. What is the pH of an aqueous solution that is 0.011 M HF ($K_a = 3.5 \times 10^{-4}$) and 0.015 M NaF?
    A) 1.95   B) 3.46   C) 3.59   D) 5.27   E) 3.33
    Ans: C

4. What is the pH of an aqueous solution that is 0.10 M HCOOH ($K_a = 1.8 \times 10^{-4}$) and 0.10 M $NaHCO_2$?
    A) 10.26   B) 3.74   C) 5.74   D) 2.38   E) 5.62
    Ans: B

5. What is the pH of an aqueous solution that is 0.20 M HNO_2 ($K_a = 4.3 \times 10^{-4}$) and 0.20 M $NaNO_2$?
    A) 3.67   B) 2.37   C) 3.37   D) 4.39   E) 10.63
    Ans: C

6. What is the pH of an aqueous solution that is 1.0 M HClO ($K_a = 3.0 \times 10^{-8}$) and 0.75 M NaClO?
    A) 7.64   B) 7.40   C) 6.36   D) 7.52   E) 6.60
    Ans: B

7. What is the pH of an aqueous solution that is 0.60 $(CH_3)_3N$ ($K_b = 6.5 \times 10^{-5}$) and 0.95 M $(CH_3)_3NHCl$?
    A) 4.39   B) 10.01   C) 3.99   D) 9.81   E) 9.61
    Ans: E

8. What is the pH of an aqueous solution that is 0.12 M $C_6H_5NH_2$ ($K_b = 4.3 \times 10^{-10}$) and 0.018 M $C_6H_5NH_3Cl$?
   A) 5.46   B) 10.19   C) 4.63   D) 3.81   E) 8.54
   Ans: A

9. What is the pH of an aqueous solution that is 0.018 M $C_6H_5NH_2$ ($K_b = 4.3 \times 10^{-10}$) and 0.12 M $C_6H_5NH_3Cl$?
   A) 5.46   B) 4.63   C) 3.81   D) 10.19   E) 8.54
   Ans: C

10. Calculate the [OH⁻] in an aqueous solution that is 0.125 M NH_3 and 0.300 M NH_4Cl. The value of K_b for NH_3 is 1.8×10^{-5}.
    A)   0.425 M
    B)   0.125 M
    C) 1.8×10^{-5} M
    D) 7.5×10^{-6} M
    E) 4.3×10^{-5} M
    Ans: D

11. Calculate the [OH⁻] in an aqueous solution that is 0.125 M NH_3 and 0.125 M NH_4Cl. The value of K_b for NH_3 is 1.8×10^{-5}.
    A) 1.8×10^{-5} M
    B) 5.5×10^{-10} M
    C) 6.7×10^{-12} M
    D)   0.125 M
    E) 1.5×10^{-3} M
    Ans: A

12. Calculate the [H⁺] in an aqueous solution that is 0.0755 M HF and 0.100 M NaF. The value of K_a for HF is 3.5×10^{-4}.
    A) 4.6×10^{-4} M
    B) 2.6×10^{-4} M
    C) 3.5×10^{-4} M
    D)   0.176 M
    E)   0.0755 M
    Ans: B

13. If 100 mL of each of the following solutions is mixed, which one produces a buffer?
    A)   1.0 M NH_3(aq) + 0.6 M KOH(aq)
    B)   1.0 M NH_4Cl(aq) + 1.0 M KOH(aq)
    C)   1.0 M NH_3(aq) + 0.4 M HCl(aq)
    D)   1.0 M NH_4Cl(aq) + 0.4 M HCl(aq)
    E)   1.0 M NH_3(aq) + 1.0 M HCl(aq)
    Ans: C

14. When the following aqueous solutions are mixed, which combination results in a buffer if the initial stoichiometric concentration of each component is 1.0 M?
    A)   NaCN + HCl
    B)   HCN + NaOH
    C)   NaCN + HCN
    D)   HNO$_3$ + HCl
    E)   HCl + NaCl
    Ans: C

15. A solution is prepared by mixing equal volumes of 0.40 M HF(aq) with 0.20 M KOH(aq). This solution is a buffer.
    Ans: True

16. A solution is prepared by mixing equal volumes of 0.40 M HF(aq) with 0.40 M KOH(aq). This solution is a buffer.
    Ans: False

17. The pH of 0.50 M HNO$_2$(aq) is 1.8. Therefore, the pH of a solution that is 0.50 M HNO$_2$(aq) and 0.10 M KNO$_2$(aq) is greater than 1.8.
    Ans: True

18. The pH of 0.30 M CH$_3$NH$_2$(aq) is 12.0. Therefore, the pH of a solution that is 0.30 M CH$_3$NH$_2$(aq) and 0.10 M CH$_3$NH$_3$Cl(aq) is greater than 12.0.
    Ans: False

19. Choose the effective pH range of an aniline-anilinium chloride buffer. The value of the K$_b$ for aniline is 4.3×10^{-10}.
    A) 3.6–5.6   B) 8.4–10.4   C) 1.1–3.1   D) 5.1–7.1   E) 10.1–12.1
    Ans: A

20. Choose the effective pH range of a HF-NaF buffer. For HF, K$_a$ = 3.5×10^{-4}.
    A) 6.0–8.0   B) 9.6–11.6   C) 5.0–7.0   D) 0.7–2.7   E) 2.5–4.5
    Ans: E

21. Choose the effective pH range of a pyridine-pyridinium chloride buffer? For pyridine, the value of K$_b$ is 1.8×10^{-9}.
    A) 9.1–11.1   B) 1.4–3.4   C) 10.3–12.3   D) 7.7–9.7   E) 4.3–6.3
    Ans: E

22. A buffer contains equal concentrations of a weak acid, HA, and its conjugate base, A$^-$. If the value of K$_a$ for HA is 1.0×10^{-9}, what is the pH of the buffer?
    A) 13.0   B) 5.0   C) 7.0   D) 1.0   E) 9.0
    Ans: E

23. A buffer contains equal concentrations of NH_3(aq) and NH_4Cl(aq). What is the pH of the buffer? (K_b (NH_3) = 1.8×10^{-5})
    A) 9.26   B) 4.74   C) 7.00   D) 13.00
    Ans: A

24. For NH_3, pK_b = 4.74. What is the pH of an aqueous buffer solution that is 0.050 M NH_3(aq) and 0.20 M NH_4Cl(aq)?
    A) 9.86   B) 5.34   C) 9.26   D) 8.66   E) 4.14
    Ans: D

25. For HF, pK_a = 3.45. What is the pH of an aqueous buffer solution that is 0.100 M HF(aq) and 0.300 M KF(aq)?
    A) 10.07   B) 2.97   C) 3.45   D) 3.93   E) 11.03
    Ans: D

26. For pyridine, pK_b = 8.75. What is the pH of an aqueous buffer solution that is 0.300 M C_5H_5N(aq) and 0.500 M C_5H_5NHCl(aq)?
    A) 8.53   B) 5.25   C) 8.97   D) 5.47   E) 5.03
    Ans: E

27. Which of the following mixtures gives a buffer with a pH greater than 7.0? For HCNO, K_a = 2.2×10^{-4} and for NH_3, K_b = 1.8×10^{-5}.
    A)   10 mL of 0.1 M NH_3(aq) + 10 mL of 0.1 M HCl(aq)
    B)   10 mL of 0.1 M HCNO(aq) + 10 mL 0f 0.1 M NaOH(aq)
    C)   10 mL of 0.1 M HCNO(aq) + 5.0 mL of 0.1 M NaOH(aq)
    D)   10 mL of 0.1 M NH_3(aq) + 10 ml of 0.1 M HCNO(aq)
    E)   10 mL of 0.1 M NH_3(aq) + 5.0 mL of 0.1 M HCl(aq)
    Ans: E

28. Which of the following mixtures gives a buffer with a pH less than 7.0? For acetic acid, K_a = 1.8×10^{-5} and for NH_3, K_b = 1.8×10^{-5}.
    A)   10 mL of 0.1 M NH_3(aq) + 10 ml of 0.1 M HCl(aq)
    B)   10 mL of 0.1 M aqueous acetic acid + 5.0 mL of 0.1 M NaOH(aq)
    C)   10 mL of 0.1 M aqueous acetic acid + 10 mL of 0.1 M NaOH(aq)
    D)   10 mL of 0.1 M aqueous acetic acid + 10 mL 0f 0.1 M NH_3(aq)
    E)   10 mL of 0.1 M NH_3(aq) + 5.0 mL of 0.1 M HCl(aq)
    Ans: B

29. If a small amount of a strong base is added to buffer made up of a weak acid, HA, and the sodium salt of its conjugate base, NaA, the pH of the buffer solution does not change appreciably because
    A)   the K_a of HA is changed.
    B)   No reaction occurs.
    C)   the strong base reacts with A^- to give HA, which is a weak acid.
    D)   the strong base reacts with HA to give AOH and H^+.
    E)   the strong base reacts with HA to give A^-, which is a weak base.
    Ans: E

30. If a small amount of a strong acid is added to buffer made up of a weak acid, HA, and the sodium salt of its conjugate base, NaA, the pH of the buffer solution does not change appreciably because
    A)   the K_a of HA is changed.
    B)   the strong acid reacts with A^- to give HA, which is a weak acid.
    C)   No reaction occurs.
    D)   the strong acid reacts with HA to give H_2A^+.
    E)   the strong acid reacts with A^- to give H_2A^+.
    Ans: B

31. The following compounds are available as 0.10 M aqueous solutions.

A) pyridine	B) $LiClO_4$	C) $HClO_4$
D) $HClO_2$	E) NaOH	F) phenol
G) triethylamine	H) HClO	I) NH_3

    Which two solutions could be used to prepare a buffer with a pH of about 7?
    Ans: E and H

32. The following compounds are available as 0.10 M aqueous solutions.

A) pyridine	B) $LiClO_4$	C) $HClO_4$
D) $HClO_2$	E) NaOH	F) phenol
G) triethylamine	H) HClO	I) NH_3

    Which two solutions could be used to prepare a buffer with a pH of about 2.5?
    Ans: D and E

33. The following compounds are available as 0.10 M aqueous solutions.

A) pyridine	B) $LiClO_4$	C) $HClO_4$
D) $HClO_2$	E) NaOH	F) phenol
G) triethylamine	H) HClO	I) NH_3

    Which two solutions could be used to prepare a buffer with a pH of about 9? More than one answer may be possible.
    Ans: E and F or C and I

34. The following compounds are available as 0.10 M aqueous solutions.

A) HCl	B) aniline	C) HCN
D) KOH	E) methylamine	F) HClO
G) HClO$_2$	H) NaClO$_2$	I) triethylamine

Pick two solutions that could be used to prepare a buffer with a pH of about 4.
Ans: A and B

35. The following compounds are available as 0.10 M aqueous solutions.

A) HCl	B) aniline	C) HCN
D) KOH	E) methylamine	F) HClO
G) HClO$_2$	H) NaClO$_2$	I) triethylamine

Pick two solutions that could be used to prepare a buffer with a pH of about 10.8. More than one answer may be possible.
Ans: A and E or A and I

36. Calculate the ratio of the molarities of CO_3^{2-} and HCO_3^- ions required to achieve buffering at pH = 9.0. For H_2CO_3, pK_{a1} = 6.37, and pK_{a2} = 10.00.
A) 0.50   B) 3.16   C) 1.65   D) 0.32   E) 0.61
Ans: D

37. Calculate the ratio of the molarities of HPO_4^{2-} and $H_2PO_4^-$ ions required to achieve buffering at pH = 7.00. For H_3PO_4, pK_{a1} = 2.12, pK_{a2} = 7.21, and pK_{a3} = 12.68.
A) 0.81   B) 1.23   C) 0.62   D) 0.21   E) 1.62
Ans: C

38. A buffer solution contains 0.0200 M acetic acid and 0.0200 M sodium acetate. What is the pH after 2.0 millimoles of HCl are added to 1.00 L of this buffer? pK_a = 4.75 for acetic acid.
A) 4.70   B) 4.84   C) 4.75   D) 4.80   E) 4.66
Ans: E

39. A buffer solution contains 0.0200 M acetic acid and 0.0200 M sodium acetate. What is the pH after 2.0 millimoles of NaOH are added to 1.00 L of this buffer? pK_a = 4.75 for acetic acid.
A) 4.75   B) 4.70   C) 4.80   D) 4.84   E) 4.66
Ans: D

40. A buffer solution contains 0.75 mol KH_2PO_4 and 0.75 mol K_2HPO_4. What is the pH after 0.10 mol KOH are added to 1.00 L of this buffer? The pK_a of $H_2PO_4^-$ is 7.21.
A) 6.91   B) 6.67   C) 7.21   D) 7.33   E) 7.09
Ans: D

41. A buffer solution contains 0.25 M $NaNO_2(aq)$ and 0.80 M $HNO_2(aq)$ ($pK_a = 3.37$).
    What is the pH after 0.10 mol HBr are added to 1.00 L of this buffer?
    A) 11.41   B) 4.15   C) 2.59   D) 9.85   E) 3.37
    Ans: C

42. Calculate the equilibrium constant for the reaction that occurs when nitric acid is added
    to the buffer HA(aq)/NaA(aq). The K_a of HA is 1.2×10^{-5}.
    Ans: 8.3×10^4

43. Calculate the equilibrium constant for the reaction that occurs when sodium hydroxide
    is added to the buffer HA(aq)/NaA(aq). The K_a of HA is 4.1×10^{-5}.
    Ans: 4.1×10^9

44. Calculate the equilibrium constant for the reaction that occurs when perchloric acid is
    added to the buffer B(aq)/BHCl(aq). The K_b of B is 3.4×10^{-5}.
    Ans: 3.4×10^9

45. Calculate the equilibrium constant for the reaction that occurs when sodium hydroxide
    is added to the buffer B(aq)/BHCl(aq). The K_b of B is 1.5×10^{-5}.
    Ans: 6.7×10^4

46. At the stoichiometric point in the titration of 0.130 M HCOOH(aq) with 0.130 M
    KOH(aq),
    A)   the pH is 7.0.   D)   the pH is greater than 7.
    B)   [HCOOH] = 0.0650 M.   E)   the pH is less than 7.
    C)   [$HCO_2–$] = 0.130 M.
    Ans: D

47. At the stoichiometric point in the titration of 0.260 M CH_3NH_2(aq) with 0.260 M
    HCl(aq),
    A)   the pH is less than 7.   D)   [CH_3NH_2] = 0.130 M.
    B)   [$CH_3NH_3^+$] = 0.260 M.   E)   the pH is greater than 7.
    C)   the pH is 7.0.
    Ans: A

48. What is the pH at the stoichiometric point for the titration of 0.100 M CH_3COOH(aq)
    with 0.100 M KOH(aq)? The value of K_a for acetic acid is 1.8×10^{-5}.
    A) 5.28   B) 8.72   C) 7.00   D) 9.26   E) 8.89
    Ans: B

49. The curve for the titration of 50.0 mL of 0.0200 M C_6H_5COOH(aq) with 0.100 M NaOH(aq) is given below. Estimate the pK_a of benzoic acid.

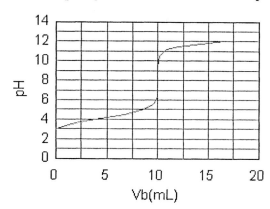

   A) 3.0   B) 4.2   C) 12.0   D) 3.8   E) 8.0
   Ans: B

50. What is the pH at the stoichiometric point for the titration of 0.26 M CH_3NH_2(aq) with 0.26 M $HClO_4$(aq)? For CH_3NH_2, $K_b = 3.6 \times 10^{-4}$.
   A) 5.72   B) 7.00   C) 5.57   D) 2.16   E) 2.01
   Ans: A

51. What is the pH at the half-stoichiometric point for the titration of 0.22 M HNO_2(aq) with 0.10 M KOH(aq)? For HNO_2, $K_a = 4.3 \times 10^{-4}$.
   A) 2.31   B) 7.00   C) 2.01   D) 3.37   E) 2.16
   Ans: D

52. What is the pH at the half-stoichiometric point for the titration of 0.88 M HNO_2(aq) with 0.10 M KOH(aq)? For HNO_2, $K_a = 4.3 \times 10^{-4}$.
   A) 3.37   B) 2.01   C) 1.86   D) 7.00   E) 1.71
   Ans: A

53. What is the concentration of acetate ion at the stoichiometric point in the titration of 0.018 M CH_3COOH(aq) with 0.036 M NaOH(aq)? For acetic acid, $K_a = 1.8 \times 10^{-5}$.
   A) 0.018 M   B) 0.0090 M   C) 0.024 M   D) 0.012 M   E) 0.036 M
   Ans: D

54. What is the concentration of acetate ion at the stoichiometric point in the titration of 0.018 M CH_3COOH(aq) with 0.072 M NaOH(aq)? For acetic acid, $K_a = 1.8 \times 10^{-5}$.
   A) 0.054 M   B) 0.036 M   C) 0.018 M   D) 0.072 M   E) 0.014 M
   Ans: E

55. For the titration of 50.0 mL of 0.020 M aqueous salicylic acid with 0.020 M KOH(aq), calculate the pH after the addition of 55.0 mL of KOH(aq). For salycylic acid, $pK_a = 2.97$.
A) 10.98   B) 7.00   C) 11.26   D) 12.02   E) 12.30
Ans: A

56. Consider the titration of 10.0 mL of 0.100 M $(CH_3)_3N(aq)$ with 0.100 M $HClO_4(aq)$. What is the formula of the main species in the solution after the addition of 10.0 mL of acid?
Ans: $CH_3)_3NH^+$

57. Consider the titration of 50.0 mL of 0.0200 M HClO(aq) with 0.100 M NaOH(aq). What is the formula of the main species in the solution after the addition of 10.0 mL of base?
Ans: ClO^-

58. Consider the titration of 50.0 mL of 0.0200 M $C_6H_5COOH(aq)$, with 0.100 M NaOH(aq). What is the formula of the main species in the solution after the addition of 10.0 mL of base? Do not consider spectator ions.
Ans: $C_6H_5COO^-$

59. Which of the following indicators would be most suitable for the titration of 0.10 M lactic acid with 0.10 M KOH(aq)? For lactic acid, $pK_a = 3.08$.
A)   methyl orange ($pK_{In} = 3.4$)   D)   bromophenol blue ($pK_{In} = 3.9$)
B)   thymol blue ($pK_{In} = 1.7$)   E)   phenol red ($pK_{In} = 7.9$)
C)   alizarin yellow ($pK_{In} = 11.2$)
Ans: E

60. Which of the following indicators would be most suitable for the titration of 0.10 M $(CH_3)_3N(aq)$ with 0.10 M $HClO_4(aq)$? For trimethyamine, $pK_b = 4.19$.
A)   bromothymol blue ($pK_{In} = 7.1$)   D)   thymol blue ($pK_{In} = 1.7$)
B)   alizarin yellow ($pK_{In} = 11.2$)   E)   phenolphthalein ($pK_{In} = 9.4$)
C)   bromocresol green ($pK_{In} = 4.7$)
Ans: C

61. The titration curve for the titration of 0.100 M H_2SO_3(aq) with 0.100 M KOH(aq) is given below.

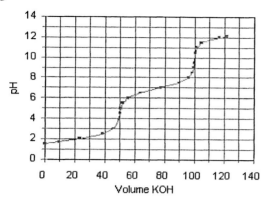

The major species in solution after 100 mL of KOH(aq) have been added are
A) H_2SO_3(aq), HSO_3^-, and Na^+(aq).
B) SO_3^{2-}(aq) and Na^+(aq).
C) HSO_3^-(aq) and Na^+(aq).
D) SO_3^{2-}(aq), OH^-(aq), and Na^+(aq).
E) HSO_3^-(aq), SO_3^{2-}(aq), and Na^+(aq).
Ans: B

62. The titration curve for the titration of 0.100 M H_2SO_3(aq) with 0.100 M KOH(aq) is given below.

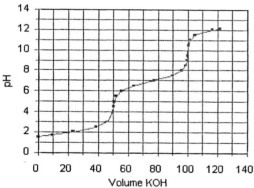

The major species in solution after 75 mL of KOH(aq) have been added are
A) HSO_3^-(aq) and Na^+(aq).
B) SO_3^{2-}(aq), and Na^+(aq).
C) SO_3^{2-}(aq), OH^-(aq), and Na^+(aq).
D) H_2SO_3(aq), HSO_3^-, and Na^+(aq).
E) HSO_3^-(aq), SO_3^{2-}(aq), and Na^+(aq).
Ans: E

63. The titration curve for the titration of 0.100 M Na_2CO_3(aq) with 0.100 M $HClO_4$(aq) is given above.

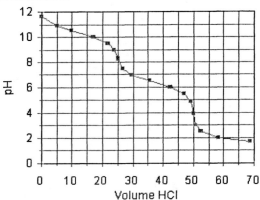

Estimate pK_{b1}.
A) 8.5   B) 6.4   C) 3.7   D) 10.3   E) 7.6
Ans: C

64. The titration curve for the titration of 0.100 M Na_2CO_3(aq) with 0.100 M $HClO_4$(aq) is given below.

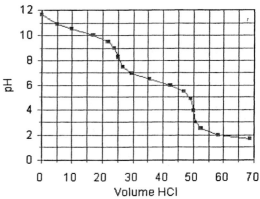

The main species in the solution after the addition of 35 mL of $HClO_4$ are
A) HCO_3^-, H_2CO_3, Na^+, and ClO_4^-.   D) CO_3^{2-}, Na^+, and ClO_4^-.
B) H_2CO_3, Na^+, and ClO_4^-.   E) HCO_3^-, Na^+, and ClO_4^-.
C) CO_3^{2-}, HCO_3-, Na^+, and ClO_4^-.
Ans: A

65. The titration curve for the titration of 0.100 M Na_2CO_3(aq) with 0.100 M $HClO_4$(aq) is given below.

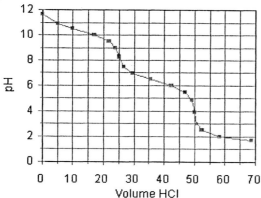

Estimate pK_{b2}.
A) 7.6   B) 10.3   C) 6.4   D) 8.5   E) 3.7
Ans: A

66. A certain weak acid has a K_a of 2.0×10^{-5}. What is the equilibrium constant for the reaction of this acid with a strong base?
Ans: 2.0×10^9

67. What is the equilibrium constant for the titration reaction involving $HClO_4$(aq) and $Ba(OH)_2$(aq)?
A) 1.0×10^{14} B) 2.0×10^{14} C) 1.0×10^7 D) 1.0×10^{-14}
Ans: A

68. The relationship between the solubility in water, s, and K_{sp} for the ionic solid M_2A_3 is
A) $K_{sp} = 108 \, s^5$ B) $K_{sp} = 5s$ C) $K_{sp} = 6s^2$ D) $K_{sp} = s^5$
Ans: A

69. Which of the following water-insoluble salts is much more soluble in 1.0 M $HClO_4$(aq)?
A) AgCl   B) $PbCO_3$ C) Hg_2Br_2 D) PbI_2 E) AgI
Ans: B

70. Silver bromide is most soluble in
A)   pure H_2O(l).
B)   dilute HNO_3(aq).
C)   0.10 M $AgNO_3$(aq).
D)   dilute NH_3(aq).
E)   0.10 M NaCl(aq).
Ans: D

71. If equal volumes of 0.004 M $Pb(NO_3)_2(aq)$ and 0.004 M KI(aq) are mixed, what reaction, if any, occurs? The value of K_{sp} for PbI_2 is 1.4×10^{-8}.
    A) The solution turns purple due to formation of I_2.
    B) $PbI_2(s)$ precipitates.
    C) $KNO_3(s)$ precipitates.
    D) No reaction occurs.
    E) K_{sp} changes to 9×10^{-9}.
    Ans: D

72. Calculate the value of the equilibrium constant for the reaction
    $$AgCl(s) + 2NH_3(aq) \rightleftharpoons Ag(NH_3)_2{}^+(aq) + Cl^-(aq)$$
    given $K_{sp} = 1.6 \times 10^{-10}$ for silver chloride and $K_f = 1.6 \times 10^7$ for the ammonia complex of Ag^+ ions, $Ag(NH_3)_2+$.
    A) 1.0×10^{-17} B) 6.3×10^9 C) 6.3×10^{-8} D) 1.0×10^{17} E) 2.6×10^{-3}
    Ans: E

73. The Cu^{2+} ion can be separated from Ag^+, Ca^{2+}, and K^+ in aqueous solution by
    A) precipitation of Cu^{2+} as CuS(s) at pH 1.
    B) precipitation of Cu^{2+} as $Cu(OH)_2(s)$ with 6 M NaOH(aq).
    C) precipitation of Cu^{2+} as $CuCl_2(s)$ with 6 M HCl(aq).
    D) precipitation of Ag^+, Ca^{2+}, and K^+ as the carbonates.
    E) None of these procedures will separate Cu^{2+} from the other ions.
    Ans: A

74. Calculate the solubility product of calcium hydroxide if the solubility of $Ca(OH)_2(s)$ in water at 25°C is 0.011 M.
    A) 1.5×10^{-8} B) 1.1×10^{-5} C) 2.7×10^{-6} D) 5.3×10^{-6} E) 1.2×10^{-4}
    Ans: D

75. If you wish to increase the solubility of silver benzoate, a preservative, you would
    A) add sodium hydroxide.   D) add sodium benzoate.
    B) decrease the pH.   E) add silver nitrate.
    C) add sodium acetate.
    Ans: B

76. You have available the following reagents as 0.10 M aqueous solutions: NaOH, HCl, HCN ($pK_a = 9.31$), aniline ($pK_b = 9.13$), HNO_2 ($pK_a = 3.25$), and CH_3NH_2 ($pK_b = 3.34$). Which two reagents would you use to make a buffer with a pH of 10.6?
    A) NaOH and HCN   C) HCl and CH_3NH_2
    B) HCl and aniline   D) NaOH and HNO_2
    Ans: C

77. What is the main factor that determines the pH of any buffer?
    Ans: The pK_a of the weak acid or conjugate acid.

78. What is the equilibrium constant for the titration reaction involving $CH_3NH_2(aq)$ and $HBr(aq)$?
    A) 1.0×10^{14} B) 2.8×10^3 C) 2.8×10^{-11} D) 3.6×10^{10}
    Ans: D

79. Consider the titration of 15.0 mL of 0.200 M $H_3PO_4(aq)$ with 0.200 M $NaOH(aq)$. What is/are the major species in solution after the addition of 15.0 mL of base?
    A) $H_2PO_4^-(aq)$ C) $PO_4^{3-}(aq)$
    B) $H_2PO_4^-(aq)$ and $HPO_4^{2-}(aq)$ D) $H_3PO_4(aq)$ and $H_2PO_4^-(aq)$
    Ans: A

80. Consider the titration of 15.0 mL of 0.200 M $H_3PO_4(aq)$ with 0.200 M $NaOH(aq)$. What is/are the major species in solution after the addition of 30.0 mL of base?
    A) $OH^-(aq)$ B) $H_3PO_4(aq)$ and $H_2PO_4^-(aq)$ C) $HPO_4^{2-}(aq)$ D) $PO_4^{3-}(aq)$
    Ans: C

81. If the molar solubility of the compound M_2A_3 is 7.0×10^{-6} M, what is the K_{sp} for this compound?
    A) 1.7×10^{-26} B) 1.8×10^{-24} C) 2.9×10^{-10} D) 3.5×10^{-5}
    Ans: B

82. The K_{sp} for mercury(I) iodide is 1.2×10^{-28}. What is the solubility of mercury(I) iodide?
    A) 3.9×10^{-10} B) 1.1×10^{-14} C) 5.2×10^{-8} D) 3.1×10^{-10}
    Ans: D

83. The solubility of silver bromide is greater in 0.10 M $NaClO_4(aq)$ than in pure water. True or False?
    Ans: False

84. The solubility of silver chloride is greater in 0.01 M $NH_3(aq)$ due to formation of the coordination complex $Ag(NH_3)_2^+(aq)$. True or False?
    Ans: True

85. When silver ions form a coordination complex with the thiosulfate ion, the latter acts as a Lewis acid. True or False?
    Ans: False

Chapter 12: Electrochemistry

1. Given: $Cr(OH)_3(s) \rightarrow CrO_4^{2-}(aq)$, basic solution.
   How many electrons appear in the balanced half-reaction?
   A) 3   B) 6   C) 5   D) 7   E) 4
   Ans: A

2. Given: $S_2O_4^{2-}(aq) \rightarrow SO_3^{2-}(aq)$, basic solution.
   How many electrons appear in the balanced half-reaction?
   A) 3   B) 6   C) 1   D) 4   E) 2
   Ans: E

3. Given: $NO_3^-(aq) \rightarrow NO(g)$, acidic solution.
   How many electrons appear in the balanced half-reaction?
   A) 8   B) 3   C) 2   D) 4   E) 6
   Ans: B

4. Given: $N_2H_5^+(aq) \rightarrow N_2(g)$.
   How many electrons appear in the balanced half-reaction?
   A) 6   B) 2   C) 1   D) 4   E) 5
   Ans: D

5. Given: $PH_3(g) \rightarrow P_4(s)$, basic solution.
   How many electrons appear in the balanced half-reaction?
   A) 12   B) 9   C) 6   D) 8   E) 3
   Ans: A

6. Given: $Zn(s) + OH^-(aq) + H_2O(l) + NO_3^-(aq) \rightarrow Zn(OH)_4^{2-}(aq) + NH_3(g)$
   If the coefficient of NO_3^- in the balanced equation is 1, how many electrons are transferred in the reaction?
   A) 10   B) 6   C) 2   D) 4   E) 8
   Ans: E

7. In the determination of iron in vitamins, Fe^{2+} is titrated with permanganate, MnO_4^-, in acidic solution. The products of the reaction are Fe^{3+} and Mn^{2+}. In the balanced equation, the number of electrons transferred is
   A) 5   B) 1   C) 10   D) 7
   Ans: A

8. For the cell diagram
$$Pt \mid H_2(g), H^+(aq) \parallel Cu^{2+}(aq) \mid Cu(s)$$
Which reaction occurs at the anode?
A) $Cu(s) \rightarrow Cu^{2+}(aq) + 2e^-$
B) $2H^+(aq) + 2e^- \rightarrow H_2(g)$
C) $2H^+(aq) + Cu(s) \rightarrow H_2(g) + Cu^{2+}(aq)$
D) $Cu^{2+}(aq) + 2e^- \rightarrow Cu(s)$
E) $H_2(g) \rightarrow 2H^+(aq) + 2e^-$
Ans: E

9. When the $Ag(s) \mid AgCl(s) \mid Cl^-(aq)$ electrode acts as a cathode, the reaction is
A) $Ag^+(aq) + e^- \rightarrow Ag(s)$.
B) $Ag(s) + Cl^-(aq) \rightarrow AgCl(s) + e^-$.
C) $Ag(s) \rightarrow Ag^+(aq) + e^-$.
D) $2AgCl(s) + 2e^- \rightarrow 2Ag^+(aq) + Cl_2(g)$.
E) $AgCl(s) + e^- \rightarrow Ag(s) + Cl^-(aq)$.
Ans: E

10. Write the cell diagram for the reaction
$$2AgCl(s) + H_2(g) \rightarrow 2Ag(s) + 2H^+(aq) + 2Cl^-(aq)$$
A) $Pt \mid Cl^-(aq) \mid H^+(aq) \parallel H_2(g) \mid AgCl(s) \mid Ag(s)$
B) $Ag(s) \mid AgCl(s) \mid Cl^-(aq) \parallel H^+(aq) \mid H_2(g) \mid Pt$
C) $Pt \mid H_2(g) \mid H^+(aq) \parallel Cl^-(aq) \mid AgCl(s) \mid Ag(s)$
D) $Pt \mid H_2(g) \mid H^+(aq) \parallel Cl^-(aq) \mid Ag(s) \mid Pt$
E) $Ag(s) \mid AgCl(s) \mid H^+(aq) \parallel Cl^-(aq) \mid H_2(g) \mid Pt$
Ans: C

11. In a working electrochemical cell (+ cell voltage), the cations in the salt bridge move toward the cathode.
Ans: True

12. In a working electrochemical cell (+ cell voltage), the electrons flow from the anode through the external circuit to the cathode.
Ans: True

13. When equilibrium is reached in an electrochemical cell, the voltage reaches its maximum value.
Ans: False

14. The standard potential of the Cu^{2+}/Cu electrode is +0.34 V and the standard potential of the cell

    $$Ag(s) | AgCl(s) | Cl^-(aq) \| Cu^{2+}(aq) | Cu(s)$$

    is +0.12 V. What is the standard potential of the $AgCl/Ag,Cl^-$ electrode?
    A) −0.46 V   B) −0.22 V   C) +0.24 V   D) +0.46 V   E) +0.22 V
    Ans: E

15. The standard potential of the Cu^{2+}/Cu electrode is +0.34 V and the standard potential of the cell

    $$Pb(s) | Pb^{2+}(aq) \| Cu^{2+}(aq) | Cu(s)$$

    is +0.47 V. What is the standard potential of the Pb^{2+}/Pb electrode?
    A) −0.26 V   B) +0.81 V   C) −0.81 V   D) −0.13 V   E) +0.13 V
    Ans: D

16. If the standard potentials for the couples Cu^{2+}/Cu, Ag^+/Ag, and Fe^{2+}/Fe are +0.34, +0.80, and −0.44 V, respectively, which is the strongest reducing agent?
    A) Fe   B) Ag   C) Ag^+ D) Cu   E) Fe^{2+}
    Ans: A

17. When a sample of an unknown metal is dropped into 1 M $H^+(aq)$ under standard conditions, bubbles are observed. The unknown metal could be silver.
    Ans: False

18. A cell that uses bromine to oxidize hydrogen to H^+ under standard conditions at 298 K has a positive potential.
    Ans: True

19. A cell that uses bromine to oxidize chloride ion under standard conditions at 298 K has a positive potential.
    Ans: False

20. If the standard potentials for the couples Fe^{3+}/Fe^{2+}, $MnO_42^-,H^+/Mn^{2+},H_2O$, Zn^{2+}/Zn, V^{3+}/V^{2+}, and Br_2/Br^- are +0.77, +1.51, −0.76, −0.26, and +1.09 V, respectively, which is the strongest oxidizing agent?
    A) Zn^{2+} B) Fe^{3+} C) Mn^{2+} D) Br_2 E) MnO_4^-
    Ans: E

21. Which species will reduce Br_2 but not V^{3+}?
    A) Ce   B) Zn   C) Cu   D) Cr^{2+} E) Al
    Ans: C

22. Which species will oxidize Cr^{2+} but not Mn^{2+}?
    A) Pb^{4+} B) O_3 in acidic medium   C) Zn^{2+} D) Fe^{2+} E) V^{3+}
    Ans: E

23. Which species will reduce Ag^+ but not Fe^{2+}?
    A) Pt   B) Au   C) V   D) Cr   E) H_2
    Ans: E

24. Which pair of metals will dissolve in nitric acid?
    A) Pt, Ag   B) Pt, Au   C) Ag, Fe   D) Ag, Au   E) Pt, Fe
    Ans: C

25. Which metal will dissolve in hydrochloric acid?
    A) All of the metals listed will dissolve.   B) Fe   C) Ag   D) Pt   E) Au
    Ans: B

26. Which of the following is the strongest reducing agent?
    A) F^- B) Co^{2+} C) Fe^{2+} D) H_2 E) Cr^{2+}
    Ans: E

27. Which of the following is the strongest oxidizing agent?
    A) O_3 B) MnO_4^- C) $Cr_2O_7^{2-}$ D) Cl_2
    Ans: A

28. Given: $Ag^+(aq) + e^- \rightarrow Ag(s)$ $E° = 0.80$ V
    $Fe^{3+}(aq) + e^- \rightarrow Fe^{2+}(aq)$ $E° = 0.77$ V
    $Cu^{2+}(aq) + 2e^- \rightarrow Cu(s)$ $E° = 0.34$ V
    Which is the strongest reducing agent?
    A) Ag   B) Cu^{2+} C) Cu   D) Ag^+ E) Fe^{2+}
    Ans: C

29. Which of the following occurs when $HNO_3(aq)$, Cu(s), and Pt(s) are mixed under standard conditions?
    A)   Pt(s) dissolves.   D)   Pt(s) dissolves and $H_2(g)$ is formed.
    B)   No reaction takes place.   E)   Cu(s) dissolves and $H_2(g)$ is formed.
    C)   Cu(s) dissolves.
    Ans: C

30. Which of the following occurs when HCl(aq), Cu(s), and Fe(s) are mixed under standard conditions?
    A) $O_2(g)$ is formed.   D)   No reaction takes place.
    B)   Cu(s) dissolves.   E)   Fe(s) dissolves.
    C) $Cl_2(g)$ is formed.
    Ans: E

31. Consider the following reaction:
$$2Ag^+(aq) + Cu(s) \rightarrow Cu^{2+}(aq) + 2Ag(s)$$
If the standard potentials of Ag^+ and Cu^{2+} are +0.80 V and +0.34 V, respectively, calculate the value of $E°$ for the given reaction.
A) +1.48 V   B) −1.26 V   C) −0.46 V   D) +1.26 V   E) +0.46 V
Ans: E

32. Consider the following reaction:
$$2Cu^+(aq) \rightarrow Cu(s) + Cu^{2+}(aq)$$
If the standard potentials of Cu^{2+} and Cu^+ are +0.34 and +0.52 V, respectively, calculate the value of $E°$ for the given reaction.
A) +0.86 V   B) +0.70 V   C) +0.18 V   D) −0.18 V   E) −0.70 V
Ans: C

33. Calculate $E°$ for the following cell.
$$Zn(s) \,|\, Zn^{2+}(aq) \,\|\, Cl^-(aq) \,|\, AgCl(s) \,|\, Ag(s)$$
A) +0.54 V   B) +1.20 V   C) −1.20 V   D) +0.98 V   E) −0.54 V
Ans: D

34. The standard potential of the cell
$$Ag(s) \,|\, AgCl(s) \,\|\, Cl^-(aq) \,|\, Cu^{2+}(aq) \,|\, Cu(s)$$
is +0.12 V at 25°C. If the standard potential of the Cu^{2+}/Cu couple is +0.34 V, calculate the standard potential of the $AgCl/Ag,Cl^-$ couple.
A) −0.12 V   B) −0.46 V   C) +0.46 V   D) −0.22 V   E) +0.22 V
Ans: E

35. If the standard free energy change for combustion of 1 mole of $CH_4(g)$ is −818 kJ·mol^{-1}, calculate the standard voltage that could be obtained from a fuel cell using this reaction.
A) −1.06 V   B) +0.53 V   C) +4.24 V   D) +8.48 V   E) +1.06 V
Ans: E

36. Consider the cell below at standard conditions:
$$Zn(s) \,|\, Zn^{2+}(aq) \,\|\, Fe^{2+}(aq) \,|\, Fe(s)$$
Calculate the value of $\Delta G_r°$ for the reaction that occurs when current is drawn from this cell.
A)   −62 kJ·mol^{-1} D)   +230 kJ·mol^{-1}
B)   −230 kJ·mol^{-1} E)   −31 kJ·mol^{-1}
C)   +62 kJ·mol^{-1}
Ans: A

37. If the standard potential for $Tl^{3+}(aq)/Tl^+(aq)$ is 1.21 V and the standard potential for $Tl^+(aq)/Tl(s)$ is -0.34 V, calculate the standard potential for
    $$Tl^{3+}(aq) + 3e^- \rightarrow Tl(s).$$
    A) 0.69 V   B) 0.87 V   C) 1.55 V   D) 0.09 V   E) 0.29 V
    Ans: A

38. If the standard potential for $Ti^{3+}(aq)/Ti^{2+}(aq)$ is -0.37 V and the standard potential for $Ti^{2+}(aq)/Ti(s)$ is -1.63 V, calculate the standard potential for
    $$Ti^{3+}(aq) + 3e^- \rightarrow Ti(s).$$
    A) -1.19 V   B) -0.40 V   C) -2.00 V   D) -1.26 V   E) -1.21 V
    Ans: E

39. If the standard potential for $Cu^{2+}(aq)/Cu^+(aq)$ is 0.15 V and the standard potential for $Cu^{2+}(aq)/Cu(s)$ is 0.34 V, calculate the standard potential for
    $$Cu^+(aq) + e^- \rightarrow Cu(s).$$
    A) $+0.32$ V   B) $+0.64$ V   C) $+0.53$ V   D) $+0.83$ V   E) $+0.49$ V
    Ans: C

40. The standard potential of the cell
    $$Pb(s)\,|\,PbSO_4(s)\,|\,SO_4^{2-}(aq)\,\|\,Pb^{2+}(aq)\,|\,Pb(s)$$
    is $+0.23$ V at 25°C. Calculate the equilibrium constant for the reaction of 1 M $Pb^{2+}(aq)$ with 1M $SO_4^{2-}(aq)$.
    A) 3.7×10^{16} B) 8.0×10^{17} C) 6.0×10^7 D) 1.7×10^{-8} E) 7.7×10^3
    Ans: C

41. The standard potential of the cell
    $$Pb(s)\,|\,PbSO_4(s)\,|\,SO_4^{2-}(aq)\,\|\,Pb^{2+}(aq)\,|\,Pb(s)$$
    is $+0.23$ V at 25°C. Calculate the K_{sp} of $PbSO_4$.
    A) 1.3×10^{-18} B) 1.7×10^{-8} C) 1.3×10^{-4} D) 2.7×10^{-17} E) 6.0×10^7
    Ans: B

42. The equilibrium constant for the reaction
    $$2Hg(l) + 2Cl^-(aq) + Ni^{2+}(aq) \rightarrow Ni(s) + Hg_2Cl_2(s)$$
    is 5.6×10^{-20} at 25°C. Calculate the value of E° for a cell utilizing this reaction.
    A) $+0.57$ V   B) -0.25 V   C) $+1.14$ V   D) -1.14 V   E) -0.57 V
    Ans: E

43. If E° for the disproportionation of $Cu^+(aq)$ to $Cu^{2+}(aq)$ and $Cu(s)$ is $+0.18$ V at 25°C, calculate the equilibrium constant for the reaction.
    A) 1.2×10^6 B) 3.9×10^{74} C) 2.5×10^{14} D) 35.7   E) 3.3×10^{12}
    Ans: A

44. The standard voltage of the cell

$$Pt \mid H_2(g) \mid H^+(aq) \parallel Cl^-(aq) \mid AgCl(s) \mid Ag(s)$$

is +0.22 V at 25°C. Calculate the equilibrium constant for the reaction below.

$$2AgCl(s) + H_2(g) \rightarrow 2Ag(s) + 2H^+(aq) + 2Cl^-(aq)$$

A) 3.7   B) 7.4   C) 5.2×10^3 D) 1.7×10^3 E) 2.7×10^7

Ans: E

45. The standard voltage of the cell

$$Ag(s) \mid AgBr(s) \mid Br^-(aq) \parallel Ag^+(aq) \mid Ag(s)$$

is +0.73 V at 25°C. Calculate the K_{sp} for AgBr.

A) 3.9×10^{-29} D) 5.1×10^{14}
B) 2.2×10^{12} E) 4.6×10^{-13}
C) 2.0×10^{-15}

Ans: E

46. The standard voltage of the cell

$$Ag(s) \mid AgBr(s) \mid Br^-(aq) \parallel Ag^+(aq) \mid Ag(s)$$

is +0.73 V at 25°C. Calculate the equilibrium constant for the cell reaction.

A) 5.1×10^{14} D) 4.6×10^{-13}
B) 2.0×10^{-15} E) 3.9×10^{-29}
C) 2.2×10^{12}

Ans: C

47. Consider the following cell:

$$Pb(s) \mid PbSO_4(s) \mid SO_4{}^{2-}(aq, 0.60\ M) \parallel H^+(aq, 0.70\ M) \mid H_2(g, 192.5\ kPa) \mid Pt$$

If E° for the cell is 0.36 V at 25°C, write the Nernst equation for the cell at this temperature.

A) $E = 0.36 + 0.01285\ln[1.90/\{(0.70)^2(0.60)\}]$
B) $E = 0.36 - 0.02569\ln[192.5/\{(0.70)^2(0.60)\}]$
C) $E = 0.36 + 0.01285\ln[192.5/\{(0.70)^2(0.60)\}]$
D) $E = 0.36 - 0.01285\ln[1.90/\{(0.70)^2(0.60)\}]$
E) $E = 0.36 - 0.01285\ln[1.90/\{(0.70)(0.60)\}]$

Ans: D

48. Consider the following cell:

$$Ag(s) \mid Ag^+(aq, 0.100\ M) \parallel Ag^+(aq, 0.100\ M) \mid Ag(s)$$

What is the voltage of this cell?

A) 0   B) +0.0592 V   C) +0.0296 V   D) +0.80 V

Ans: A

49. Consider the following cell:

    $Pt \mid Fe^{2+}(aq, 0.50 \text{ M}), Fe^{3+}(aq, 0.30 \text{ M}) \parallel Fe^{3+}(aq, 0.035 \text{ M}), Fe^{2+}(aq, 0.010) \mid Pt$

    The standard potential for the Fe^{3+}/Fe^{2+} couple is +0.77 V. Calculate the cell voltage at 25°C.

    Ans: +0.045 V

50. Calculate E for the half-reaction below.

    $2H^+(aq, 1.00 \times 10^{-5} \text{ M}) + 2e^- \rightarrow H_2(g, 1.00 \text{ atm})$

    A) 0 V   B) +0.592 V   C) −0.592 V   D) −0.296 V   E) +0.296 V

    Ans: D

51. Calculate E for the half-reaction below.

    $2H^+(aq, 1.00 \text{ M}) + 2e^- \rightarrow H_2(g, 1.00 \text{ atm})$

    Ans: 0 V

52. Consider the following cell:

    $Pt \mid H_2(g, 1 \text{ atm}) \mid H^+(aq, ? \text{ M}) \parallel Ag^+(aq, 1.0 \text{ M}) \mid Ag(s)$

    If the voltage of this cell is 1.04 V at 25°C and the standard potential of the Ag^+/Ag couple is +0.80 V, calculate the hydrogen ion concentration in the anode compartment.

    A) 4.6×10^{-10} M   D) 1.0 M
    B) 8.8×10^{-5} M   E) 3.7×10^{-8} M
    C) 9.4×10^{-3} M

    Ans: B

53. Consider the following cell:

    $Zn(s) \mid Zn^{2+}(aq, 0.100 \text{ M}) \parallel Cl^-(aq, ? \text{ M}) \mid Cl_2(g, 0.500 \text{ atm}) \mid Pt$

    For this cell, E° = 2.12 V and E = 2.27 V at 25°C. Calculate the $Cl^-(aq)$ concentration in the cathode compartment.

    A) 1.2×10^{-1} M   D) 1.5×10^{-3} M
    B) 4.3×10^{-5} M   E) 6.5×10^{-3} M
    C) 2.9×10^{-3} M

    Ans: E

54. Consider the following cell:

    $Zn(s) \mid Zn^{2+}(aq, 0.200 \text{ M}) \parallel H^+(aq, ?) \mid H_2(g, 1.00 \text{ atm}) \mid Pt$

    If E = +0.66 V and E° = +0.76 V at 25°C, calculate the concentration of H^+ in the cathode cell compartment.

    A) 2.1×10^{-2} M   D) 9.2×10^{-3} M
    B) 4.0×10^{-1} M   E) 4.0×10^{-3} M
    C) 8.4×10^{-5} M

    Ans: D

Use the following to answer questions 55-58:

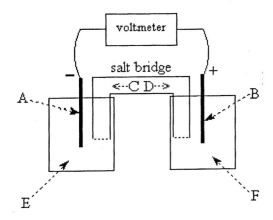

55. The galvanic cell shown above uses the half-cells Mg^{2+}/Mg and Zn^{2+}/Zn, and a salt bridge containing KCl(aq). The voltmeter gives a positive voltage reading. Identify A and write the half-reaction that occurs in that compartment. Does the size of the electrode A increase or decrease during operation of the cell? What is the voltmeter reading?
Ans: A is Mg(s); $Mg(s) \rightarrow Mg^{2+}(aq) + 2e^-$; A decreases in size; +1.60 V

56. The galvanic cell shown above uses the half-cells Pb^{2+}/Pb and Zn^{2+}/Zn, and a salt bridge containing KCl(aq). The voltmeter gives a positive voltage reading. Identify A and write the half-reaction that occurs in that compartment. Does the size of the electrode A increase or decrease during operation of the cell? What is the voltmeter reading?
Ans: A is Zn(s); $Zn(s) \rightarrow Zn^{2+}(aq) + 2e^-$; A decreases in size; +0.63 V

57. The galvanic cell shown above uses the half-cells Pb^{2+}/Pb and Zn^{2+}/Zn, and a salt bridge containing KCl(aq). The voltmeter gives a positive voltage reading. The electrode B could be inert platinum metal or lead.
Ans: True

58. The galvanic cell shown above uses the half-cells Pb^{2+}/Pb and Zn^{2+}/Zn, and a salt bridge containing KCl(aq). The voltmeter gives a positive voltage reading. The electrode B could be inert platinum metal or zinc.
Ans: False

Use the following to answer questions 59-64:

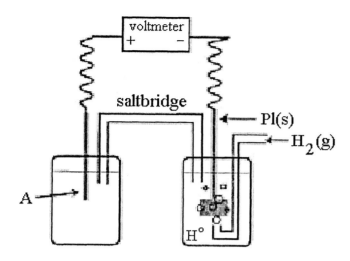

59. In the cell shown above, A is a standard Zn^{2+}/Zn electrode connected to a standard hydrogen electrode (SHE). If the voltmeter reading is −0.76 V, which half-reaction occurs in the left-hand cell compartment?
    A) $Zn^{2+}(aq) + 2e^- \rightarrow Zn(s)$ B) $Zn(s) \rightarrow Zn^{2+}(aq) + 2e^-$
    Ans: B

60. In the cell shown above, A is a standard Zn^{2+}/Zn electrode connected to a standard hydrogen electrode (SHE). If the voltmeter reading is −0.76 V, what is the equation for the cell reaction?
    A) $Zn^{2+}(aq) + H_2(g) \rightarrow Zn(s) + 2H^+(aq)$ B) $Zn(s) + 2H^+(aq) \rightarrow Zn^{2+}(aq) + H_2(g)$
    Ans: B

61. In the cell shown above, A is a standard Ag^+/Ag electrode connected to a standard hydrogen electrode (SHE). If the voltmeter reading is +0.80 V, which half-reaction occurs in the left-hand cell compartment?
    A) $Ag(s) \rightarrow Ag^+(aq) + e^-$ B) $Ag^+(aq) + e^- \rightarrow Ag(s)$
    Ans: B

62. In the cell shown above, A is a standard Ag^+/Ag electrode connected to a standard hydrogen electrode (SHE). If the voltmeter reading is +0.80 V, what is the equation for the cell reaction?
    A) $Ag(s) + H^+(aq) \rightarrow Ag^+(aq) + ½H_2(g)$
    B) $Ag^+(aq) + ½H_2(g) \rightarrow Ag(s) + H^+(aq)$
    Ans: B

63. In the cell shown above, A is a standard Zn^{2+}/Zn electrode connected to a standard hydrogen electrode (SHE). If the voltmeter reading is −0.76 V, which electrode is negative?
   Ans: A

64. In the cell shown above, A is a standard Ag^+/Ag electrode connected to a standard hydrogen electrode (SHE). If the voltmeter reading is +0.80 V, which electrode is negative?
   Ans: SHE

65. Which metal would be suitable to provide cathodic protection from corrosion for an iron bridge?
   A) Cu   B) Ni   C) None of the metals listed is suitable.   D) Sn   E) Pb
   Ans: C

66. The products of the electrolysis of $CuSO_4(aq)$ are
   A) $H_2(g)$ and $H_2SO_3(aq)$.
   B) $H_2SO_3(aq)$ and $O_2(g)$.
   C) $Cu(s)$ and $H_2SO_3(aq)$.
   D) $Cu(s)$ and $O_2(g)$.
   E) $H_2(g)$ and $O_2(g)$.
   Ans: D

67. Sodium is produced by electrolysis of molten sodium chloride. What are the products at the anode and cathode, respectively?
   A) $Na(l)$ and $O_2(g)$
   B) $Cl^-(aq)$ and $Na_2O(l)$
   C) $Cl_2(g)$ and $Na_2O(l)$
   D) $O_2(g)$ and $Na(l)$
   E) $Cl_2(g)$ and $Na(l)$
   Ans: E

68. How many moles of $O_2(g)$ are produced by electrolysis of $Na_2SO_4(aq)$ if 0.120 A is passed through the solution for 65.0 min?
   A)   0.00121 mol
   B)   0.0000808 mol
   C)   0.00242 mol
   D)   0.00485 mol
   E)   0.0000202 mol
   Ans: A

69. If 8686 C of charge is passed through molten magnesium chloride, calculate the number of moles of $Mg(l)$ produced.
   A)   0.0225 mol
   B)   0.0450 mol
   C)   2.00 mol
   D)   0.0110 mol
   E)   0.0900 mol
   Ans: B

70. How long will it take to deposit 0.00235 moles of gold by electrolysis of $KAuCl_4(aq)$ using a current of 0.214 amperes?
A) 17.7 min   B) 26.5 min   C) 70.7 min   D) 106 min   E) 53.0 min
Ans: E

71. How many moles of $Cl_2(g)$ are produced by the electrolysis of concentrated sodium chloride if 2.00 A are passed through the solution for 4.00 hours? The equation for this process, the "chloralkali" process, is
$$2NaCl(aq) + 2H_2O(l) \rightarrow 2NaOH(aq) + H_2(g) + Cl_2(g)$$
A)   0.0745 mol   D)   0.00248 mol
B)   0.149 mol   E)   0.298 mol
C)   0.447 mol
Ans: B

72. The half-reaction that occurs at cathode when 1 M $AgNO_3(aq)$ is electrolyzed is

_____.

Ans: Ag(s)

73. When a lead-acid battery discharges, sulfuric acid is produced. True or False?
Ans: False

74. What current is required to produce 91.6 g of chromium meta from chromium(VI) oxide in 12.4 hours?
Ans: 22.8 A

75. How many seconds are required to produce 4.99 mg of chromium metal from an acidic solution of potassium dichromate, using a current of 0.234 A?
Ans: 237 s

76. In order to convert hydrazine, N_2H_4, to nitric acid,
A) an oxidizing agent is required.   B) a reducing agent is required.
Ans: A

77. In the Daniell cell, the anode is zinc metal and the cathode is copper metal. When the cell operates, the anode gets smaller and the cathode gets larger. True or False?
Ans: True

78. For the reduction of Cu^{2+} by Zn, $\Delta G^\circ = -212$ kJ/mol and $E^\circ = +1.10$ V. If the coefficients in the chemical equation for this reaction are multiplied by 2, $\Delta G^\circ = -424$ kJ/mol. This means $E^\circ = +2.20$ V. True or False?
Ans: False

79. When KI(aq) is electrolyzed at a concentration of 1 M, the product at the anode is I_2. True or False?
Ans: False

80. Galvanized iron is protected from corrosion because zinc reduces any Fe^{2+} formed. True or False?
    Ans: True

81. For the cell diagram
    $$Pt \mid H_2(g) \mid H^+(aq) \parallel Co^{3+}(aq), Co^{2+}(aq) \mid Pt$$
    write the reaction that occurs at the cathode.
    Ans: $Co^{3+}(aq) + e^- \rightarrow Co^{2+}(aq)$

82. For the cell diagram
    $$Cd(s) \mid CdSO_4(aq) \mid Hg_2SO_4 \mid Hg(l)$$
    write the reaction that occurs at the cathode.
    Ans: $Hg_2SO_4(s) + 2e^- \rightarrow 2Hg(l) + SO_4^{2-}(aq)$

83. All galvanic cells require a salt bridge in order to operate. True or False?
    Ans: False

84. Consider the following cell:
    $$Zn(s) \mid Zn^{2+}(aq, 0.10\ M) \parallel Cu^{2+}(aq, 0.10\ M) \mid Cu(s)$$
    At equilibrium, what is the concentration of $Zn^{2+}(aq)$?
    Ans: 0.20 M

85. Consider the following cell:
    $$Zn(s) \mid Zn^{2+}(aq, 0.10\ M) \parallel Cu^{2+}(aq, 0.10\ M) \mid Cu(s)$$
    At equilibrium, what is the concentration of $Cu^{2+}(aq)$?
    Ans: 1.3×10^{-38} M

Chapter 13: Chemical Kinetics

1. If the average rate of decomposition of $PH_3(g)$ is 3.2 (mol PH_3)·L^{-1}·min^{-1} for the reaction $2PH_3(g) \rightarrow 2P(g) + 3H_2(g)$, the unique average reaction rate is
   A) 3.2 mol·L^{-1}·min^{-1}.
   D) 4.8 mol·L^{-1}·min^{-1}.
   B) 2.1 mol·L^{-1}·min^{-1}.
   E) 6.4 mol·L^{-1}·min^{-1}.
   C) 1.6 mol·L^{-1}·min^{-1}.
   Ans: C

2. If the average rate of formation of $H_2(g)$ is 3.90 (mol H_2)·L^{-1}·s^{-1} for the reaction $2PH_3(g) \rightarrow 2P(g) + 3H_2(g)$, the unique average reaction rate is
   A) 3.90 mol·L^{-1}·s^{-1}.
   D) 7.80 mol·L^{-1}·s^{-1}.
   B) 1.30 mol·L^{-1}·s^{-1}.
   E) 11.7 mol·L^{-1}·s^{-1}.
   C) 2.60 mol·L^{-1}·s^{-1}.
   Ans: B

3. The rate of formation of oxygen in the reaction
   $$2N_2O_5(g) \rightarrow 4NO_2(g) + O_2(g)$$
   is 2.28 (mol O_2)·L^{-1}·s^{-1}. What is the rate of formation of NO_2?
   A) 0.57 (mol NO_2)·L^{-1}·s^{-1}
   D) 1.14 (mol NO_2)·L^{-1}·s^{-1}
   B) 9.12 (mol NO_2)·L^{-1}·s^{-1}
   E) 4.56 (mol NO_2)·L^{-1}·s^{-1}
   C) 2.28 (mol NO_2)·L^{-1}·s^{-1}
   Ans: B

4. The rate of formation of $NO_2(g)$ in the reaction
   $$2N_2O_5(g) \rightarrow 4NO_2(g) + O_2(g)$$
   is 5.78 (mol NO_2)·L^{-1}·s^{-1}. What is the rate at which N_2O_5 decomposes?
   A) 0.723 (mol N_2O_5)·L^{-1}·s^{-1}
   D) 11.6 (mol N_2O_5)·L^{-1}·s^{-1}
   B) 1.45 (mol N_2O_5)·L^{-1}·s^{-1}
   E) 5.78 (mol N_2O_5)·L^{-1}·s^{-1}
   C) 2.89 (mol N_2O_5)·L^{-1}·s^{-1}
   Ans: C

5. It is important to distinguish between the reaction rate and the rate constant. The units of reaction rate are M^{-1}·s^{-1}. True or False?
   Ans: False

6. Given:
   $$2NO_2(g) + F_2(g) \rightarrow 2NO_2F(g) \qquad rate = -\Delta[F_2]/\Delta t$$
   The rate of the reaction can also be expressed as
   A) $-2\Delta[NO_2]/\Delta t$.
   D) $-\Delta[NO_2]/\Delta t$.
   B) $\Delta[NO_2F]/\Delta t$.
   E) $\frac{1}{2}\Delta[NO_2F]/\Delta t$.
   C) $\frac{1}{2}\Delta[NO_2]/\Delta t$.
   Ans: E

7. Given:
   $$4Fe^{2+}(aq) + O_2(aq) + 2H_2O(l) \rightarrow 4Fe^{3+}(aq) + 4OH^-(aq) \qquad rate = k[Fe^{2+}][OH^-]^2[O_2]$$
   The overall order of the reaction and the order with respect to O_2 are
   A) 4 and 1.   B) 5 and 1.   C) 3 and 1.   D) 4 and 2.   E) 7 and 1.
   Ans: A

8. The reaction
   $$2NO(g) + 2H_2(g) \rightarrow N_2(g) + 2H_2O(g)$$
   is first-order in H_2 and second-order in NO. Starting with equal concentrations of H_2 and NO, the rate after 25% of the H_2 has reacted is what percent of the initial rate?
   A) 75.0%   B) 42.2%   C) 6.25%   D) 56.3%   E) 1.56%
   Ans: B

9. If the rate of reaction increases by a factor of 9.6 when the concentration of reactant increases by a factor of 3.1, the order of the reaction with respect to this reactant is
   A) 1.5   B) 3   C) 4   D) 2   E) 1
   Ans: D

10. Given:
    $$2O_3(g) \rightarrow 3O_2(g) \qquad rate = k[O_3]^2[O_2]^{-1}$$
    The overall order of the reaction and the order with respect to $[O_3]$ are
    A) -1 and 3.   B) 1 and 2.   C) 0 and 1.   D) 2 and 2.   E) 3 and 2.
    Ans: B

11. Given:
    $$2A(g) + B(g) \rightarrow C(g) + D(g)$$
    When $[A] = [B] = 0.10$ M, the rate is 2.0 M·s^{-1}; for $[A] = [B] = 0.20$, the rate is 8.0 M·s^{-1}; and for $[A] = 0.10$ M, $[B] = 0.20$ M, the rate is 2.0 M·s^{-1}. The rate law is
    A)   rate = k[A].
    B)   rate = k[B]^2.
    C)   rate = k[A][B]^0.
    D)   rate = k[A][B].
    E)   rate = k[A]^2.
    Ans: E

12. If the rate of a reaction increases by a factor of 64 when the concentration of reactant increases by a factor of 4, the order of the reaction with respect to this reactant is
    A) 16.   B) 2.   C) 4.   D) 1.   E) 3.
    Ans: E

13. For the reaction

$$2A + B \rightarrow \text{products}$$

determine the rate law for the reaction given the following data:

Initial concentration, M		Initial rate, M·s^{-1}
A	B	
0.10	0.10	2.0×10^{-2}
0.20	0.10	8.0×10^{-2}
0.30	0.10	1.8×10^{-1}
0.20	0.20	8.0×10^{-2}
0.30	0.30	1.8×10^{-1}

A) rate = $k[B]^2$
B) rate = $k[A][B]^0$
C) rate = $k[A]^2$
D) rate = $k[A][B]$
E) rate = $k[A]$

Ans: C

14. If the rate of reaction increases by a factor of 2.3 when the concentration of reactant increases by a factor of 1.5, the order of the reaction with respect to this reactant is
A) 2   B) 1   C) 1.5   D) 4   E) 3
Ans: A

15. For the reaction $S_2O_8^{2-}(aq) + 3I^-(aq) \rightarrow 2SO_4^{2-}(aq) + I_3-(aq)$, rate = $k[S_2O_8^{2-}][I^-]$. When the reaction is followed under pseudo-first-order conditions with $[S_2O_8^{2-}] = 200$ mM and $[I^-] = 1.5$ mM, the rate constant was 1.82 s^{-1}. The second order rate constant, k, for the reaction is
A) 1.21×10^3 M^{-1}·s^{-1}.
B) 6.07×10^3 M^{-1}·s^{-1}.
C) 9.10 M^{-1}·s^{-1}.
D) 1.37×10^{-2} M^{-1}·s^{-1}.
E) 1.82 M^{-1}·s^{-1}.

Ans: C

16. The reaction

$$2NO(g) + 2H_2(g) \rightarrow N_2(g) + 2H_2O(g)$$

is first-order in H_2 and second-order in NO. Starting with equal concentrations of H_2 and NO, the rate after 50% of the H_2 has reacted is what percent of the initial rate?
A) 18.8%   B) 37.5%   C) 25.0%   D) 50.0%   E) 12.5%
Ans: E

17. For the reaction
$$2NO(g) + 2H_2(g) \rightarrow N_2(g) + 2H_2O(g)$$
the following data were collected.

[NO(g)]	[H2(g)]	Rate, M·s^{-1}
0.10	0.10	0.0050
0.10	0.20	0.010
0.10	0.30	0.015
0.20	0.10	0.020
0.20	0.20	0.040

What is the rate law for this reaction?
Ans: rate = $k[H_2][NO]^2$

18. The concentration–time dependence for a first-order reaction is given below.

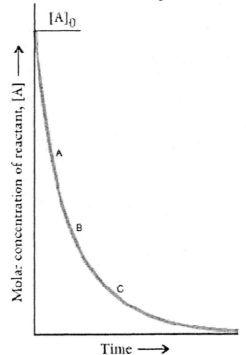

At which point on the curve is the reaction fastest?
A) A   B) B   C) C   D) A + $t_{1/2}$ E) The rates are the same at all points.
Ans: A

19. The concentration–time dependence is shown below for two first-order reactions. Which reaction has the largest rate constant?

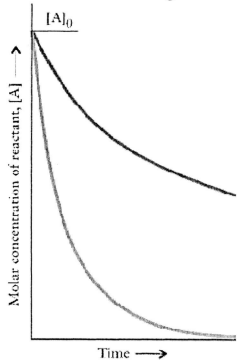

Ans: the reaction that gives the lower curve

20. The concentration–time dependence is shown below for two first order reactions. Which reaction has the greatest $t_{1/2}$?

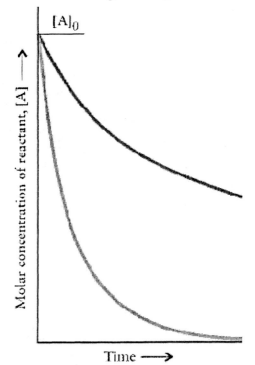

Ans: the reaction that gives the upper curve

21. The concentration–time curves are shown below for two sets of reactions, A/B and C/D. Which set of reactions has the largest rate constant?

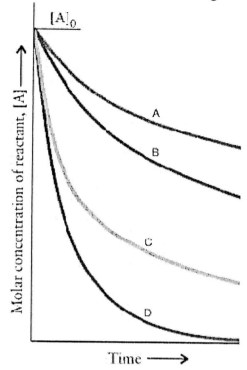

Ans: C/D

22. The concentration–time curves are shown below for two sets of reactions, A/B and C/D.

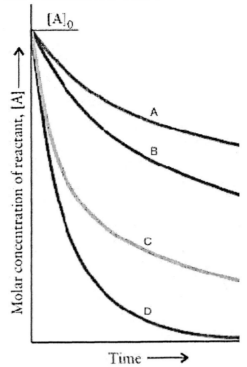

Which of the reactions are first-order?
A) B and D   B) A and B   C) C and D   D) A and C
Ans: A

23. A first-order reaction has a rate constant of 0.00300 s^{-1}. The time required for 60% reaction is
A) 153 s.   B) 73.9 s.   C) 170 s.   D) 133 s.   E) 305 s.
Ans: E

24. For a first-order reaction, after 2.00 min, 20% of the reactants remain. Calculate the rate constant for the reaction.
A) 0.0134 s^{-1} D) 0.00582 s^{-1}
B) 0.000808 s^{-1} E) 0.00186 s^{-1}
C) 74.6 s^{-1}
Ans: A

25. Consider the following reaction:
$$2N_2O(g) \rightarrow 2N_2(g) + O_2(g) \qquad \text{rate} = k[N_2O]$$
For an initial concentration of N_2O of 0.50 M, calculate the concentration of N_2O remaining after 2.0 min if $k = 3.4 \times 10^{-3} \text{ s}^{-1}$.
A) 0.50 M   B) 0.55 M   C) 0.66 M   D) 0.33 M   E) 0.17 M
Ans: D

26. A first-order reaction has a rate constant of $0.00300 \ s^{-1}$. The time required for 85%
    reaction is
    A) 632 s.   B) 23.5 s.   C) 275 s.   D) 316 s.   E) 54.2 s.
    Ans: A

27. A non-steroidal anti-inflammatory drug is metabolized with a first-order rate constant of
    $3.25 \ day^{-1}$. What is the half-life for the metabolism reaction?
    A) 0.213 day   B) 0.308 day   C) 1.63 day   D) 2.25 day
    Ans: A

28. Consider the following reaction:
    $$2N_2O(g) \rightarrow 2N_2(g) + O_2(g) \qquad rate = k[N_2O]$$
    Calculate the time required for the concentration of $N_2O(g)$ to decrease from 0.75 M to
    0.33 M. The rate constant for the reaction is $k = 6.8 \times 10^{-3} \ s^{-1}$.
    A) 2.7 min   B) 1.7 min   C) 0.87 min   D) 0.92 min   E) 2.0 min
    Ans: E

29. For a first-order reaction, after 230 s, 33% of the reactants remain. Calculate the rate
    constant for the reaction.
    A) $207 \ s^{-1}$
    B) $0.00174 \ s^{-1}$
    C) $0.00209 \ s^{-1}$
    D) $0.000756 \ s^{-1}$
    E) $0.00482 \ s^{-1}$
    Ans: E

30. Consider the following reaction:
    $$2N_2O_5(g) \rightarrow 4NO_2(g) + O_2(g) \qquad rate = k[N_2O_5]$$
    If the initial concentration of N_2O_5 is 0.80 M, the concentration after 5 half-lives is
    A) 0.11 M.   B) 0.025 M.   C) 0.032 M.   D) 0.16 M.   E) 0.050 M.
    Ans: B

31. A compound decomposes with a half-life of 8.0 s and the half-life is independent of the
    concentration. How long does it take for the concentration to decrease to one-ninth of its
    initial value?
    A) 32 s   B) 25 s   C) 72 s   D) 64 s   E) 3.6 s
    Ans: B

32. Consider the following reaction:
    $$2N_2O_5(g) \rightarrow 4NO_2(g) + O_2(g) \qquad rate = k[N_2O_5]$$
    Calculate the time for the concentration of N_2O_5 to fall to one-fourth its initial value if
    the half-life is 133 s.
    A) 133 s   B) 33.3 s   C) 266 s   D) 66.6 s   E) 533 s
    Ans: C

33. Technetium-99m, used to image the heart and brain, has a half-life of 6.00 h. What fraction of technetium-99m remains in the body after 1 day?
A) 0.0625   B) 0.250   C) 0.0313   D) 0.125
Ans: A

34. The reaction
$$2ClO_2(g) + F_2(g) \rightarrow 2FClO_2(g)$$
is first-order in both ClO_2 and F_2. When the initial concentrations of ClO_2 and F_2 are equal, the rate after 25% of the F_2 has reacted is what percent of the initial rate?
A) 75.0%   B) 18.8%   C) 37.5%   D) 28.1%   E) 12.5%
Ans: C

35. A first-order reaction has a half-life of 1.10 s. If the initial concentration of reactant is 0.384 M, how long will it take for the reactant concentration to reach 0.00100 M?
A) 9.45 s   B) 0.106 s   C) 4.10 s   D) 1.52 s   E) 0.244 s
Ans: A

36. Given: $A \rightarrow P$ rate = k[A]
If 20% of A reacts in 5.12 min, calculate the time required for 90% of A to react.
A) 52.8 min   B) 1.05 min   C) 2.42 min   D) 3170 min   E) 22.9 min
Ans: A

37. For the reaction
cyclopropane $\rightarrow$ propene
a plot of ln[cyclopropane] vs time in seconds gives a straight line with slope -4.1×10^{-3} s^{-1} at 550°C. What is the rate constant for this reaction?
A) 3.9×10^{-2} s^{-1} D) 1.8×10^{-3} s^{-1}
B) 8.2×10^{-3} s^{-1} E) 2.1×10^{-3} s^{-1}
C) 4.1×10^{-3} s^{-1}
Ans: C

38. For the reaction A $\rightarrow$ Products, the following data were collected.

time, s	0	1	2	3	4
[A], M	1.00	0.430	0.270	0.200	0.160

Determine the order of the reaction and calculate the rate constant.
Ans: second-order, 1.33 M^{-1}·s^{-1}

39. For a second-order reaction, a straight line is obtained from a plot of ln{1/[A]} vs t.
Ans: False

40. For the reaction cyclopropane(g) $\rightarrow$ propene(g) at 500°C, a plot of ln[cyclopropane] vs t gives a straight line with a slope of −0.00067 s^{-1}. What is the order of this reaction and what is the rate constant?
Ans: first-order, 6.7×10^{-4} s^{-1}

41. For a second-order reaction, a straight line is obtained from a plot of
    A)   1/[A] vs t.   D)   ln[A] vs t.
    B)   ln(1/t) vs [A].   E)   ln(t) vs [A].
    C)   [A] vs t.
    Ans: A

42. What is the half-life of a reaction that has a rate constant of 280 s^{-1}?
    A)   194 s
    B)   3.6 ms
    C)   404 ms
    D)   cannot calculate because the concentration of reactant is not given
    E)   2.5 ms
    Ans: E

43. What is the rate constant for a first-order reaction with a half-life of 9.0 ms?
    A) 13 s^{-1} B) 77 s^{-1} C) 9.0 s^{-1} D) 6.2 s^{-1} E) 0.11 s^{-1}
    Ans: B

44. For the reaction cyclobutane(g) $\rightarrow$ 2ethylene(g) at 800 K, a plot of ln[cyclobutane] vs t gives a straight line with a slope of -1.6 s^{-1}. Calculate the time needed for the concentration of cyclobutane to fall to 1/16 of its initial value.
    A) 2.3 s   B) 1.7 s   C) 1.3 s   D) 0.63 s   E) 1.6 s
    Ans: B

45. For the reaction cyclobutane(g) $\rightarrow$ 2ethylene(g) at 800 K, the half-life is 0.43 s. Calculate the time needed for the concentration of cyclobutane to fall to 1/64 of its initial value.
    A) 2.2 s   B) 2.6 s   C) 16 ms   D) 0.38 s   E) 0.43 s
    Ans: B

46. A second-order reaction has a rate constant of 1.25 M^{-1}·s^{-1}. If the initial reactant concentration is 1.0 M, calculate the time required for 90% reaction.
    A) 1.3 s   B) 7.2 s   C) 0.13 s   D) 17 s   E) 0.89 s
    Ans: B

47. For the dimerization of butadiene(g) at a certain temperature, a plot of [butadiene(g)]$^{-1}$ vs t gives a straight line with a slope of $+0.704$ M^{-1}·s^{-1}. What is the rate constant for this reaction?
    Ans: 0.704 M^{-1}·s^{-1}

48. Consider the dimerization reaction below:

$$2A \rightarrow A_2 \qquad rate = k[A]^2$$

When the initial concentration of A is 2.0 M, it requires 30 min for 60% of A to react. Calculate the rate constant.

A) $1.1 \times 10^{-3} \, M^{-1} \cdot s^{-1}$ D) $1.9 \times 10^{-4} \, M^{-1} \cdot s^{-1}$

B) $3.2 \times 10^{-4} \, M^{-1} \cdot s^{-1}$ E) $4.2 \times 10^{-4} \, M^{-1} \cdot s^{-1}$

C) $5.0 \times 10^{-4} \, M^{-1} \cdot s^{-1}$

Ans: E

49. Consider the following reaction:

$$NOBr(g) \rightarrow NO(g) + \frac{1}{2}Br_2(g)$$

A plot of $[NOBr]^{-1}$ vs time gives a straight line with a slope of $2.00 \, M^{-1} \cdot s^{-1}$. The order of the reaction and the rate constant, respectively, are

A) second-order and $0.500 \, M^{-1} \cdot s^{-1}$.   D) first-order and $0.241 \, s^{-1}$.

B) first-order and $2.00 \, s^{-1}$.   E) second-order and $16.6 \, M^{-1} \cdot s^{-1}$.

C) second-order and $2.00 \, M^{-1} \cdot s^{-1}$.

Ans: C

50. A reaction has $k = 8.39 \, M^{-1} \cdot s^{-1}$. How long does it take for the reactant concentration to drop from 0.0840 M to 0.0220 M?

A) 5.42 s   B) 2.00 s   C) 1.42 s   D) 8.39 s   E) 4.00 s

Ans: E

51. What is the half-life of a second order reaction?

Ans: $t_{1/2} = 1/k[A]_o$

52. The activation energy of a reaction is given by

A) +(slope of a plot of lnk vs 1/T) ÷ R.   D) +(slope of a plot of lnk vs 1/T) × R.

B) −(slope of a plot of lnk vs 1/T) ÷ R.   E) −(slope of a plot of lnk vs 1/T) × R.

C) −R ÷ (slope of a plot of lnk vs 1/T).

Ans: E

53. For the reaction

$$HO(g) + H_2(g) \rightarrow H_2O(g) + H(g)$$

a plot of lnk versus 1/T gives a straight line with a slope equal to -5.1×10^3 K. What is the activation energy for the reaction?

A) $42 \, kJ \cdot mol^{-1}$ D) $5.1 \, kJ \cdot mol^{-1}$

B) $98 \, kJ \cdot mol^{-1}$ E) $12 \, kJ \cdot mol^{-1}$

C) $0.61 \, kJ \cdot mol^{-1}$

Ans: A

54. A certain reaction has a rate constant of 8.8 s^{-1} at 298 K and 140 s^{-1} at 323 K. What is the activation energy for this reaction?
   A) 38 $kJ \cdot mol^{-1}$
   D) 23 $kJ \cdot mol^{-1}$
   B) 89 $kJ \cdot mol^{-1}$
   E) 1.2 $kJ \cdot mol^{-1}$
   C) 120 $kJ \cdot mol^{-1}$
   Ans: B

55. An elementary process has an activation energy of 92 kJ/mol. If the enthalpy change for the reaction is -62 kJ/mol, what is the activation energy for the reverse reaction?
   A) 154 kJ/mol   B) 62 kJ/mol   C) 92 kJ/mol   D) 30 kJ/mol
   Ans: A

56. A reaction that has a very high activation energy
   A) has a rate that is very sensitive to temperature.
   B) has a rate that does not change much with temperature.
   C) must be second-order.
   D) gives a curved Arrhenius plot.
   E) must be first-order.
   Ans: A

57. A catalyst facilitates a reaction by
   A) increasing the activation energy for the reverse reaction.
   B) lowering the activation energy of the reaction.
   C) shifting the position of the equilibrium of the reaction.
   D) decreasing the temperature at which the reaction will proceed spontaneously.
   E) making the reaction more exothermic.
   Ans: B

58. Given:
$$CH_4(g) + Cl_2(g) \rightarrow CH_3Cl(g) + HCl(g)$$
   The rate law for this elementary process is
   A) rate = $k[Cl_2]$.
   D) rate = $k[CH_4]$.
   B) $k[CH_3Cl][HCl]$.
   E) rate = $k[CH_4]^2$.
   C) rate = $k[CH_4][Cl_2]$.
   Ans: C

59. The reaction $2NO(g) + O_2(g) \rightarrow 2NO_2(g)$ has $\Delta H_r° = -114$ kJ·mol^{-1}. A possible mechanism for this reaction is given below.

$$2NO(g) \rightleftharpoons N_2O_2(g) \qquad \text{rapid equilibrium, K}$$
$$N_2O_2(g) + O_2(g) \rightarrow 2NO_2(g) \quad \text{slow, k}$$

Draw the reaction profile diagram (plot of energy vs reaction coordinate) for this reaction and label any intermediates and activated complexes.

Ans:

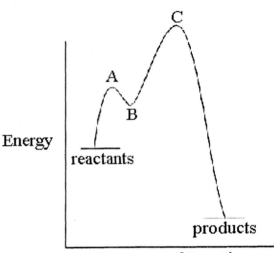

A and C are activated complexes and B is the intermediate $N_2O_2(g)$.

60. In a certain solvent, reaction (1) occurs in one step whereas reaction two occurs in two steps.

(1) $CH_3Cl + OH^- \rightarrow CH_3OH + Cl^-$

(2) $(CH_3)_3CBr + OH^- \rightarrow (CH_3)_3COH + Br^-$ Rate = $k[(CH_3)_3CBr]$

Postulate a mechanism for each reaction.

Ans:

(1) $CH_3Cl + OH^- \rightarrow [HO\text{----}CH_3\text{----}Cl] \rightarrow CH_3OH + Cl^-$

(2) $(CH_3)_3CBr \rightarrow (CH_3)_3C^+ + Br^-$ k, slow

     $(CH_3)_3C^+ + OH^- \rightarrow (CH_3)_3COH$ fast

61. The reaction $[(CN)_5CoOH_2]^{2-}(aq) + SCN^-(aq) \rightarrow [(CN)_5CoSCN]^{3-} + H_2O(l)$ has the rate law, rate = $k[(CN)_5CoOH_2^{2-}]$. Postulate a mechanism for this reaction.

Ans:

$[(CN)_5CoOH_2]^{2-}(aq) \rightarrow [(CN)_5Co]^{2-}(aq) + H_2O(l)$ k, slow

$[(CN)_5Co]^{2-}(aq) + SCN^-(aq) \rightarrow [(CN)_5CoSCN]^{3-}$ fast

62. The reaction profile for the following reaction
    $[(CN)_5CoOH_2]^{2-}(aq) + SCN^-(aq) \rightarrow [(CN)_5CoSCN]^{3-} + H_2O(l)$
    is

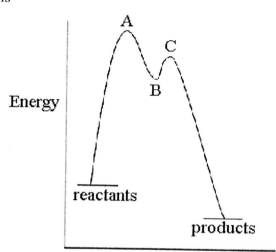

    Identify the structure of B. What do A and C represent?
    Ans: $B = [(CN)_5Co]^{2-}$; A and C are activated complexes.

63. The rate law for the following mechanism is
    $NO_2(g) + F_2(g) \rightarrow NO_2F(g) + F(g)$ k_1, slow
    $F(g) + NO_2(g) \rightarrow NO_2F(g)$ k_2, fast
    A)   rate = $k_2[NO_2]^2$.
    B)   rate = $k_2[NO_2][F]$.
    C)   rate = $k_1[NO_2][F_2]$.
    D)   rate = $k_1k_2[NO_2]^2$.
    E)   rate = $k_1[NO_2F][F]$.
    Ans: C

64. The rate law for the following mechanism is
    $ClO^-(aq) + H_2O(l) \rightleftharpoons HOCl(aq) + OH^-(aq)$ K, fast
    $I^-(aq) + HOCl(aq) \rightarrow HOI(aq) + Cl^-(aq)$ k_1, slow
    $HOI(aq) + OH^-(aq) \rightarrow OI^-(aq) + H_2O(l)$ k_2, fast
    A)   rate = $k_1[I^-][HOCl]$.
    B)   rate = $k_1K[ClO^-][I^-][OH^-]$.
    C)   rate = $k_1K[ClO^-][I^-][OH^-]^{-1}$.
    D)   rate = $k_1k_2K[ClO^-][I^-]$.
    E)   rate = $k_1K[ClO^-][I^-]$.
    Ans: C

65. The reaction

$$2NO_2(g) \rightarrow 2NO(g) + O_2(g)$$

is postulated to occur via the mechanism below:

$$NO_2(g) + NO_2(g) \rightarrow NO(g) + NO_3(g) \qquad \text{slow}$$
$$NO_3(g) \rightarrow NO(g) + O_2(g) \qquad \text{fast}$$

An intermediate in this reaction is

A) $NO_2(g)$.   B) $ON\text{-}NO_3(g)$.   C) $O_2(g)$.   D) $NO(g)$.   E) $NO_3(g)$.

Ans: E

66. The reaction between nitrogen dioxide and carbon monoxide is thought to occur by the following mechanism:

$$2NO_2(g) \rightarrow NO_3(g) + NO(g) \qquad k_1, \text{slow}$$
$$NO_3(g) + CO(g) \rightarrow NO_2(g) + CO_2(g) \qquad k_2, \text{fast}$$

The rate law for this mechanism is

A)   rate = $k_1k_2[NO_2]^2[CO]$.   D)   rate = $k_2[NO_3][CO]$.

B)   rate = $(k_1/k_2)[NO_2]^2[CO]$.   E)   rate = $k_1[NO_2]^2$.

C)   rate = $k_1[NO_3][NO]$.

Ans: E

67. The rate law for the mechanism below is

$$Cl_2(g) \rightleftharpoons 2Cl(g) \qquad K_1, \text{fast}$$
$$Cl(g) + CO(g) \rightleftharpoons COCl(g) \qquad K_2, \text{fast}$$
$$COCl(g) + Cl_2(g) \rightarrow COCl_2(g) + Cl(g) \qquad k_3, \text{slow}$$

A)   rate = $k_3K_10.5K_2[CO][Cl_2]^{1.5}$.   D)   rate = $k_3[COCl][Cl_2]$.

B)   rate = $k_3K_10.5K_2[CO][Cl_2]^{0.5}$.   E)   rate = $k_3[COCl][Cl_2]^{1.5}$.

C)   rate = $k_3K_1K_2[CO][Cl_2]$.

Ans: A

68. A possible mechanism for the reaction $2NO(g) + O_2(g) \rightarrow 2NO_2(g)$ is given below.

$$2NO(g) \rightleftharpoons N_2O_2(g) \qquad k_1, k_1', \text{fast}$$
$$N_2O_2(g) + O_2(g) \rightarrow 2NO_2(g) \qquad k_2, \text{slow}$$

Application of the steady-state approximation gives

A) $[N_2O_2] = 0$.

B) $k_1[NO]^2 - k_2[N_2O_2][O_2] = 0$.

C) $[N_2O_2] = (k_1/k_2)[NO]^2$.

D) $k_1[NO]^2 - k_1'[N_2O_2] - k_2[N_2O_2][O_2] = 0$.

E) $[N_2O_2] = (k_1/k_1')[NO]^2$.

Ans: D

69. The HBr synthesis is thought to involve the following reactions:

$$Br_2 \rightarrow 2Br\cdot \qquad (1)$$
$$Br\cdot + H_2 \rightarrow HBr + H\cdot \qquad (2)$$
$$H\cdot + Br_2 \rightarrow HBr + Br\cdot \qquad (3)$$
$$2Br\cdot \rightarrow Br_2 \qquad (4)$$
$$2H\cdot \rightarrow H_2 \qquad (5)$$
$$H\cdot + Br\cdot \rightarrow HBr \qquad (6)$$

The chain termination reactions in this mechanism are reactions
A) 3, 4, and 5.   B) 4 and 5.   C) 4, 5, and 6.   D) 6.   E) 3 and 4.
Ans: C

70. The HBr synthesis is thought to involve the following reactions:

$$Br_2 \rightarrow 2Br\cdot \qquad (1)$$
$$Br\cdot + H_2 \rightarrow HBr + H\cdot \qquad (2)$$
$$H\cdot + Br_2 \rightarrow HBr + Br\cdot \qquad (3)$$
$$2Br\cdot \rightarrow Br_2 \qquad (4)$$
$$2H\cdot \rightarrow H_2 \qquad (5)$$
$$H\cdot + Br\cdot \rightarrow HBr \qquad (6)$$

The chain propagation reactions in this mechanism are reactions
A) 6.   B) 1, 2, 3, and 6.   C) 2 and 3.   D) 1, 2, and 3.   E) 2, 3, and 6.
Ans: C

71. In the Michaelis-Menten mechanism of enzyme reaction, the Michaelis constant, K_M, is
A) $K_M = k_1/k_2$.
B) $K_M = k_1'/k_2$.
C) $K_M = (k_1' + k_2)/k_1$.
D) $K_M = k_1' + k_2$.
E) $K_M = k_1/k_1'$.
Ans: C

72. The reaction profile for the mechanism
$$NO_2(g) + F_2(g) \rightarrow NO_2F(g) + F(g) \qquad slow$$
$$F(g) + NO_2(g) \rightarrow NO_2F(g) \qquad fast$$
shows
A)   two maxima, the first maximum being highest.
B)   two maxima, the second maximum being highest.
C)   one maximum for the second step.
D)   two maxima, both the same height.
Ans: A

73. The reaction $2NO(g) + O_2(g) \rightarrow 2NO_2(g)$ has $\Delta H_r° = -114$ kJ·mol^{-1}. A possible mechanism for this reaction is given below.

$2NO(g) \rightleftharpoons N_2O_2(g)$ rapid equilibrium, K

$N_2O_2(g) + O_2(g) \rightarrow 2NO_2(g)$ slow, k

If the activation energy for the reverse reaction is 225 kJ·mol^{-1}, what is the activation energy for the forward reaction?

A) Because an intermediate is involved, the forward activation energy cannot be calculated with the given information.
B) 111 kJ·mol^{-1}
C) 339 kJ·mol^{-1}
D) 114 kJ·mol^{-1}
E) 55.5 kJ·mol^{-1}

Ans: B

74. A possible mechanism for the reaction $A + B + D \rightarrow E + F$ is

$A + B \rightleftharpoons C$ k_1, k_{-1}, fast

$C + D \rightarrow E + F$ k_2

where C is a reactive intermediate present in very low concentrations. Calculate the steady state concentration of C in terms of the rate constants for the individual steps and the concentrations of the reactants.

Ans: $[C] = k_1[A][B]/\{k_{-1} + k_2[D]\}$

75. The reaction $2A + B \rightarrow D + E$ has the rate law, rate $= a[A]^2[B]/(b + c[A])$ where a, b, and c are constants. The following mechanism has been proposed for this reaction.

$A + B \rightleftharpoons I$ k_1, k_{-1}

$I + A \rightarrow D + E$ k_2

I is an unstable intermediate present in minute concentrations. Show that this mechanism leads to the observed rate law and evaluate the constants a, b, and c in terms of the rate constants k_1, k_{-1}, and k_2.

Ans: $a = k_1k_2$, $b = k_{-1}$, $c = k_2$ or $a = k_1$, $b = k_{-1}/k_2$, $c = 1$.

76. For a zero-order reaction, the rate constant has the same units as the rate of reaction. True or False?

Ans: True

77. The order of a reaction is always a whole number. True or False?

Ans: False

78. The reduction of M^{3+} by Cr^{2+} was studied with $[Cr^{2+}]$ 100 times the concentration of Cr^{2+}. When $[Cr^{2+}] = 0.0050$ M, the rate was 2.5 s^{-1}. The rate constant for this reaction is

A) 1.3 M^{-1}·s^{-1} B) 0.013 M^{-1}·s^{-1} C) 5.0 M^{-1}·s^{-1} D) 500 M^{-1}·s^{-1}

Ans: D

79. Given: $A + B \rightarrow P$ rate = $k[A][B]$
    Which of the following is true?
    A) $k = \ln 2 / t_{1/2}$ C) $[A]_t = [A]_o / (1 + kt[A]_o)$
    B) $\ln[A]_t = -kt + \ln[A]_o$ D) $1/[A]_t = 1/kt$
    Ans: C

80. The rate law for a reaction can be determined from the coefficients in the overall reaction. True or False?
    Ans: False

81. For the reaction $A \rightarrow$ Products, the following data were collected.

time,s	0	1.00	2.00	3.00	4.00
$[A]_t$, M	1.00	0.430	0.270	0.200	0.160

    the half-life for this reaction.
    A) 0.521 s   B) 0.752 s   C) 0.922 s   D) 1.08 s
    Ans: B

82. For the elementary reaction $A + 2B \rightarrow$ Products, rate = $k[A][B]$. True or False?
    Ans: False

83. A mechanism for the decomposition of ozone, $2O_3 \rightarrow 3O_2$, is given below.
    $O_3 \rightarrow O_2 + O$
    $O + O_3 \rightarrow O_2 + O_2$
    The molecularity for the first and second elementary reactions is ____ and ____.
    Ans: 1, 2

84. The fraction of molecules that collide with a kinetic energy equal to the activation energy for a reaction decreases rapidly with an increase in temperature. True or False?
    Ans: False

85. When a catalyst is used to increase the rate of a given reaction, the rate law is changed. True or False?
    Ans: True

Chapter 14: The Elements: The First Four Main Groups

1. The valence electrons of elements in the lower left of the periodic table experience the greatest effective nuclear charge.
    Ans: False

2. Circle the species that has the largest radius: Ca^{2+}, S^{2-}, S, Cl.
    Ans: S^{2-}

3. Write an equation for the second ionization energy of calcium.
    Ans: $Ca^+(g) \rightarrow Ca^{2+}(g) + e^-$

4. The electron affinities of elements are always positive.
    Ans: False

5. Trends in electron affinity are similar to those for ionization energy.
    Ans: True

6. Which of the following has the smallest polarizability?
    A) Ca^{2+} B) S^{2-} C) Cl   D) S
    Ans: A

7. Which of the following compounds is predicted to have bonds with the most covalent character?
    A) MgF_2 B) LiF   C) CsI   D) MgI_2 E) $BaBr_2$
    Ans: D

8. Which of the following has the largest atomic radius?
    A) In   B) Ga   C) Ge   D) Al
    Ans: A

9. Which of the following exists mainly as cations in compounds?
    A) P   B) Ga   C) As   D) Si   E) Sr
    Ans: E

10. Which of the following pairs of elements have similar polarizability?
    A) Be and Al   B) Li and Na   C) B and Al   D) Be and Mg   E) Mg and B
    Ans: A

11. All of the following form saline hydrides except
    A) barium.   B) beryllium.   C) lithium.   D) magnesium.   E) calcium.
    Ans: B

12. Write the formula of the hydride ion.
    Ans: H^-

13. When metallic hydrides are heated, they
    A)   decompose explosively to yield metal oxides.
    B)   liberate oxygen and hydrogen.
    C)   liberate hydrogen.
    D)   produce hydrogen ions.
    E)   do not change because they are extremely stable.
    Ans: C

14. All of the oxides of the following elements are amphoteric except
    A) P   B) Ga   C) As   D) Sb
    Ans: A

15. All of the following can act as Lewis acids except
    A) NO.   B) SO_3.   C) Na_2O.   D) NO_2.   E) CO_2.
    Ans: C

16. Aluminum oxide dissolves in aqueous base. What are the products of this reaction?
    A) $Al(OH_2)_6^{3+}$(aq) and H_2(g)   D) $Al(OH)_4^-$(aq) and H_2(g)
    B) Al_2O_3(s) and H_2(g)   E) $Al(OH)_4^-$(aq) and O_2(g)
    C) $Al(OH_2)_6^{3+}$(aq) and O_2(g)
    Ans: D

17. Al_2O_3(s) dissolves in aqueous acid to produce
    A) $Al(OH_2)_6^{3+}$(aq).   D) $Al(OH_2)_6^{3+}$(aq) and H_2(g).
    B) $Al(OH)_4^-$(aq) and O_2(g).   E) $Al(OH)_4^-$(aq) and H_2(g).
    C) $Al(OH)_4^-$(aq).
    Ans: A

18. Which of the following pairs of elements both have acidic oxides?
    A) Ca and As   B) Mg and Al   C) B and Li   D) In and Sn   E) P and Se
    Ans: E

19. Which of the following is a basic oxide?
    A) CaO(s)   B) CO_2(g)   C) SO_2(g)   D) CO(g)   E) SiO_2(s)
    Ans: A

20. Which of the following is an acidic oxide?
    A) CaO(s)   B) MgO(s)   C) Bi_2O_3(s)   D) Na_2O(s)   E) SiO_2(s)
    Ans: E

21. Hydrogen can be made from fossil fuels in a series of two catalyzed reactions. One of these reactions, the **re-forming reaction**, is
    A) $Cu(s) + 2H^+(aq) \rightarrow Cu^{2+}(aq) + H_2(g)$.
    B) $2H_2O(l) \rightarrow 2H_2(g) + O_2(g)$.
    C) $CH_3OH(l) \rightarrow 2H_2(g) + CO(g)$.
    D) $CH_4(g) + H_2O(g) \rightarrow CO(g) + 3H_2(g)$.
    E) $Zn(s) + 2HCl(aq) \rightarrow ZnCl_2(aq) + H_2(g)$.
    Ans: D

22. Use a table of standard potentials to determine the voltage for the water-splitting reaction to produce hydrogen and oxygen from water.
    Ans: -1.23 V

23. Which of the following can be extracted by hydrometallurgical reduction of their ions with hydrogen? Consult a table of standard potentials.
    A) Ni   B) Fe   C) Ag   D) Zn   E) Pb
    Ans: C

24. Which of the following metals cannot be extracted by hydrometallurgical reduction of their ions with hydrogen? Ag, Zn, Cu, Ni, Au, Pt, Sn
    A)   Au, Pt   D)   Zn, Ni, Sn
    B)   Ag, Au, Pt   E)   Ag, Cu, Au, Pt
    C)   Zn, Ni, Au
    Ans: D

25. The equation for the hydrometallurgical extraction of silver is
    A) $Ag^+(aq) + Cl^-(aq) \rightarrow AgCl(s)$.
    B) $Ag^+(aq) + 2S_2O_3^{2-}(aq) \rightarrow Ag(S_2O_3)_2^{3-}(aq)$.
    C) $Ag^+(aq) + 2CN^-(aq) \rightarrow Ag(CN)_2^-(aq)$.
    D) $2Ag(s) + 2H^+(aq) \rightarrow Ag^+(aq) + H_2(g)$.
    E) $Ag^+(aq) + H_2(g) \rightarrow 2Ag(s) + 2H^+(aq)$.
    Ans: E

26. When saline hydrides dissolve in water they
    A)   produce $O_2(g)$.
    B)   reduce water and produce $H_2(g)$.
    C)   Saline hydrides are insoluble in water.
    D)   reduce water and produce O_3.
    E)   produce $H_3O^+(aq)$.
    Ans: B

27. Order the following ions according to their radii from largest to smallest: H^-, Cl^-, F^-.
    Ans: $Cl^- > H^- > F^-$

28. The alkali metals all react with water. Which is the most reactive metal?
    A) Li   B) Na   C) Cs   D) Rb   E) K
    Ans: C

29. Both the hydride ion and sodium metal reduce water.
    Ans: True

30. What property of lithium makes the lithium-ion battery an excellent energy storage device?
    A) Li^+ has a strong polarizing power.
    B)   Lithium-ion is a good reducing agent.
    C)   Lithium has a high density.
    D)   Lithium has the most negative standard potential of all the elements.
    E) Li^+ has a small radius.
    Ans: D

31. All of the following react with water to form hydrogen and a hydroxide except
    A) Mg.   B) Na.   C) Ca.   D) K.   E) Be.
    Ans: E

32. The reaction of potassium superoxide with water vapor produces
    A) $K_2O(s)$.   B) $H_2O_2(l)$.   C) $KHCO_3(s)$.   D) $O_2(g)$.   E) $H_2(g)$.
    Ans: D

33. Lithium and sodium burn in air to produce
    A) $Li_2O(s)$ and $Na_2O(s)$.   D) $LiO_2(s)$ and $Na_2O_2(s)$.
    B) $LiO_2(s)$ and $Na_2O(s)$.   E) $Li_2O_2(s)$ and $NaO_2(s)$.
    C) $Li_2O(s)$ and $Na_2O_2(s)$.
    Ans: C

34. What is the formula of anhydrous "Epsom salts"?
    A) $Mg(OH)_2$ B) $MgCl_2$ C) $MgSO_4$ D) $CaCl_2$ E) $CaSO_4$
    Ans: C

35. What occurs when beryllium is added to aqueous sodium hydroxide?
    A) $Be(OH)_2(s)$ precipitates.
    B)   Beryllium dissolves to form $[Be(OH)_4]^{2-}(aq)$.
    C)   Beryllium dissolves to form $[Be(OH_2)_6]^{2+}(aq)$.
    D) $O_2(g)$ is produced.
    E)   No reaction occurs because beryllium is not soluble in aqueous sodium hydroxide.
    Ans: B

36. What is the formula of "quicklime"?
    A) CaO   B) $MgSO_4$ C) $Ca(OH)_2$ D) $CaCO_3$ E) $CaSO_4$
    Ans: A

37. What is the formula of "baking soda"?
    A) $Na_2CO_3 \cdot 10H_2O$ B) $NaHCO_3$ C) $NaHSO_4$ D) Na_2SO_4 E) Na_2CO_3
    Ans: B

38. Quicklime, CaO(s), reacts with C(s) to produce A plus carbon monoxide. When A is added to water, B and C are produced. Write the formulas of A, B, and C.
    A)   Ca(s), O_2(g), and Ca(OH)$_2$(aq)   D)   Ca(s), H_2(g), and Ca(OH)$_2$(aq)
    B)   CaO(s) does not react with C(s).   E) CaC_2(s), C_2H_2(g), and Ca(OH)$_2$(aq)
    C) CaC_2(s), H_2(g), and Ca(OH)$_2$(aq)
    Ans: E

39. In the following reaction
    $$2KNO_3(s) + 4C(s) \rightarrow K_2CO_3(s) + 3CO(g) + N_2(g)$$
    the reducing agent is
    A) CO(g).   B) KNO_3(s).   C) K_2CO_3(s).   D) C(s).   E) N_2(g).
    Ans: D

40. Which of the substances below can be used to extinguish a magnesium fire?
    A)   sand
    B)   carbon dioxide
    C)   nitrogen
    D)   All of the substances listed react with burning magnesium.
    E)   water
    Ans: D

41. Aluminum metal is produced by
    A)   reduction of Al_2O_3 with carbon.
    B)   electrolysis of a molten mixture of alumina, Al_2O_3, and cryolite, Na_3AlF_6.
    C)   treatment of Al_2O_3 with sodium hydroxide.
    D)   the thermite reaction.
    E)   electrolysis of brine containing $Al_2(SO_4)_3$.
    Ans: B

42. In the thermite reaction, aluminum is the reducing agent.
    Ans: True

43. The formula of boric acid written as a Lewis acid is
    A) $B(OH)_3$.   B) $HB(OH)_3$.   C) $HB(OH)_4$.   D) $B(OH)_4-$.   E) $HB(OH)_2$.
    Ans: A

44. The formula of boric acid written as a Brønsted acid is
    A) $HB(OH)_2$.   D) $HB(OH)_3$.
    B) $B(OH_2)_3{}^{3+}$.   E) $B(OH)_3OH_2$.
    C) $B(OH)_4{}^-$.
    Ans: E

45. When magnesium burns in air, the product(s) is(are)
    A)   MgO.
    B)   MgO and Mg_3N_2.
    C)   MgAr.
    D) $MgCO_3$.
    E)   MgS and MgO.
    Ans: B

46. Boron nitride is produced from the reaction of
    A)   B(s) and HCN(g).
    B)   B(s) and NH_3(g).
    C) B_2O_3(s) and NH_3(g).
    D) B_2H_6(g) and N_2(g).
    E)   B(s) and N_2(g).
    Ans: B

47. The role of magnesium in chlorophyll is
    A)   a hydrogenation catalyst.
    B)   to keep the ring rigid.
    C)   a reducing agent.
    D)   an oxidizing agent.
    E)   to react with water to produce hydrogen.
    Ans: B

48. Sodium borohydride is produced from the reaction of
    A) $B(OH)_3$(aq) and BCl_3(l).
    B)   NaH(s) and B(s).
    C)   NaH(s) and B_2O_3(s).
    D)   NaH(s) and $B(OH)_3$(aq).
    E)   NaH(s) and BCl_3(l).
    Ans: E

49. The formula of diborane is
    A) BH_3.   B) B_2H_6.   C) B_2H_4.   D) B_2H_2.   E) B_2H.
    Ans: B

50. Diborane is produced by the reaction of
    A)   B(s) and NH_3(g).
    B) BH_3(g) and B(s).
    C)   B(s) and H_2(g).
    D) BF_3(g) and $NaBH_4$(s).
    E) BCl_3(g) and NaH(s).
    Ans: D

51. The equation that corresponds to the K_a for boric acid is
    A) $B_4O_5(OH)_4{}^{2-}$(aq) + $5H_2O$(l) $\rightleftharpoons$ $4H_3BO_3$(aq) + $2OH^-$(aq).
    B) H_3BO_3(aq) + H_2O(l) $\rightleftharpoons$ $HB(OH)_3{}^+$(aq) + OH^-(aq).
    C) $(OH)_3BOH_2$(aq) + H_2O(l) $\rightleftharpoons$ $B(OH)_4{}^-$(aq) + H_3O^+(aq).
    D) $B(OH)_4{}^-$(aq) + H^+(aq) $\rightleftharpoons$ $B(OH)_3$(aq) + H_2O(l).
    E) $B_4O_7{}^{2-}$(aq) + H_2O(l) $\rightleftharpoons$ $B_4O_8H^{3-}$(aq) + H^+(aq).
    Ans: C

52. Diborane has
    A)   2 bridging hydrogens and 4 terminal hydrogens.
    B)   1 BF_3^- anion and 1 BF_3^+ cation.
    C)   4 bridging hydrogens and 2 terminal hydrogens.
    D)   6 terminal hydrogens and 1 boron-boron bond.
    E)   1 bridging hydrogen, 4 terminal hydrogens, and an ionic hydrogen.
    Ans: A

53. Sodium borohydride is an oxidizing agent.
    Ans: False

54. The molecule Al_2Br_6 is formed from two $AlBr_3$ molecules. This reaction is
    A)   a precipitation reaction.
    B)   a reduction reaction.
    C)   an oxidation reaction.
    D)   redox dimerization reaction.
    E)   a Lewis acid-base complex formation reaction.
    Ans: E

55. When aluminum sulfate is dissolved in water, the resulting solution is acidic or basic? Explain.
    Ans: acidic; $Al(OH_2)_6^{3+}(aq) + H_2O(l) \rightleftharpoons Al(OH)(H_2O)_5^{2+}(aq) + H_3O^+(aq)$.

56. The oxides of In and Tl
    A)   are amphoteric.   D)   are unstable and cannot be isolated.
    B)   react with base.   E)   are basic.
    C)   react with water.
    Ans: E

57. Which of the following are allotropes?
    A)   graphite, diamond, and C_{60}
    B)   silicon, carbon, and C_{60}
    C)   boron carbide and carbon
    D)   carbon monoxide and carbon dioxide
    E)   silicon carbide, diamond, and C_{60}
    Ans: A

58. The allotropes of carbon are graphite, diamond, and the fullerenes. These are all network solids.
    Ans: False

59. Which of the following is not true?
    A)   In diamond, each carbon atom is sp^3-hybridized and linked to its four neighbors, with all electrons in C-C σ-bonds.
    B) C_{60} is molecular and thus soluble in solvents like benzene.
    C)   Graphite consists of planar sheets of sp^2-hybridized carbon atoms and electrons can move from one carbon to another through a delocalized π-network formed by the overlap of unhybridized p-orbitals.
    D)   Graphite is an electrically conducting solid.
    E)   Diamond is an excellent conductor of heat.
    Ans: C

60. Which of the following concentrated acids can be safely transported in passivated lead containers?
    A) $HClO_4$ B) H_2SO_4 C) HI   D) HCl   E) HNO_3
    Ans: B

61. Lead is obtained from the oxide PbO by
    A)   reduction with aluminum.
    B)   treatment with concentrated sulfuric acid.
    C)   reduction with carbon.
    D)   reduction with CO(g).
    E)   heating to 1600°C.
    Ans: C

62. All of the following reactions involve carbon monoxide acting as a reducing agent except
    A) $Ni(s) + 4CO(g) \rightarrow Ni(CO)_4(g)$.
    B) $PbO(s) + CO(g) \rightarrow Pb(s) + CO_2(g)$.
    C) $2CO(g) + O_2(g) \rightarrow 2CO_2(g)$.
    D) $CO(g) + H_2O(g) \rightarrow CO_2(g) + H_2(g)$.
    E) $Fe_2O_3(s) + 3CO(g) \rightarrow 2Fe(s) + 3CO_2(g)$.
    Ans: A

63. Carbon dioxide is the formal anhydride of
    A) H_2CO_3 B) HCOOH   C) H_2CO D) CO_3^{2-}
    Ans: A

64. The sublimation of *dry ice* is spontaneous at 25°C and 1 atm. What is the sign of ΔH° and ΔS°, respectively?
    A) 0, +   B) +,−   C) −,+   D) +, +   E) −,−
    Ans: D

65. Which of the following molecules would require the most energy to split into gaseous atoms?
    A) CO   B) F_2 C) O_2 D) N_2 E) Cl_2
    Ans: A

66. The mineral spondumene contains long, straight chain silicates involving the SiO_3^{2-} unit. What is the shape of SiO_3^{2-}?
    A)   T-shaped   D)   trigonal pyramidal
    B)   tetrahedral   E)   see-saw
    C)   trigonal planar
    Ans: C

67. Which of the following is a saline carbide?
    A) SiC   B) CaC_2 C) Mo_2C D) Fe_3C E) W_2C
    Ans: B

68. Which of the following is a covalent carbide?
    A) W_2C B) Fe_3C C) SrC_2 D) CaC_2 E) SiC
    Ans: E

69. Which of the following is an interstitial carbide?
    A) SrC_2 B) Al_4C_3 C) CaC_2 D) K_2C_2 E) W_2C
    Ans: E

70. Saline carbides contain the anion(s)
    A) C_2^{2-} and C^{4-}.   D) CO_3^{2-} and HCO_3^-.
    B) CH_2^{2-}.   E) CN^- and N^{3-}.
    C) HC_2-.
    Ans: A

71. The products of the reaction of $Al_4C_3(s)$ with water are
    A) $Al_2O_3(s)$ and $CH_4(g)$.   D) $Al_2O_3(s)$ and $C_2H_2(g)$.
    B) $Al(OH)_3(s)$ and $C_2H_2(g)$.   E) $Al(OH)_3(s)$ and $CH_4(g)$.
    C) $Al(OH)_3(s)$ and $C_2H_4(g)$.
    Ans: E

72. All of the following contain silicates in various forms except
    A) asbestos.   B) mica.   C) talc.   D) molecular sieves.   E) alum.
    Ans: E

73. Consider the following possible reactions:

    1. $B(OH)_3(aq) + H_2O(l) \rightarrow$
    2. $SiH_4(g) + H_2O(l)$ (trace $OH^-(aq)$) $\rightarrow$
    3. $CCl_4(l) + H_2O(l) \rightarrow$
    4. $SiCl_4(l) + H_2O(l) \rightarrow$

    Which of the reactions that occur are Lewis acid–base reactions?
    A) (3) and (4)   B) (2)   C) (1), (3), and (4)   D) (1)   E) (1) and (4)
    Ans: E

74. Silane is produced in the following reactions.

    a. $SiCl_4 + LiAlH_4 \rightarrow SiH_4 + LiAlCl_4$
    b. $Mg_2Si + 4HCl \rightarrow SiH_4 + 2MgCl_2$

    Classify each reaction as a redox or displacement reaction.
    Ans: (a) displacement; (b) redox.

75. All of the following are silicon-oxygen compounds except
    A) asbestos.   B) quartz.   C) talc.   D) diamond.   E) mica.
    Ans: D

76. Polarizability measures the ease with which an electron cloud can be distorted and is greatest for electron-rich atoms like fluorine. True or False?
    Ans: False

77. The compound BeI_2 is an ionic compound. True or False?
    Ans: False

78. Which of the following compounds does not exist?
    A) NCl_3 B) PCl_5 C) PCl_3 D) NCl_5
    Ans: D

79. All of the following elements form hydrides except
    A) K   B) Be   C) Ca   D) Rb
    Ans: B

80. Which of the following statements is false?
    A)   Oxides of nonmetals are acidic.
    B)   Elements with intermediate ionization energies form amphoteric oxides.
    C)   Elements with low ionization energies commonly form ionic oxides.
    D)   Oxides of main-group metals are acidic.
    Ans: D

81. Solid sodium reduces nitrogen gas to form sodium nitride, Na_3N. True or False?
    Ans: False

82. Lithium and magnesium have similar properties and their compounds have covalent character. True or False?
    Ans: True

83. Beryllium and aluminum have similar properties because
    A)   both have similar electron affinities.
    B)   both have almost identical ionization energies.
    C)   both ions are highly polarizing due to their small size and high charge.
    D)   both have virtually identical atomic radii.
    Ans: C

84. Which Group 13 element is most likely to have an oxidation number of +1 in its compounds?
    Ans: Tl

85. Both boric acid and nickel(II) chloride are weak acids in aqueous solutions. Explain.
    Ans: Both form a Lewis acid/Lewis base complex with water and this complex is a weak acid.

Chapter 15: The Elements: The Last Four Main Groups

1. The nitrate ion can only act as an oxidizing agent in redox reactions.
   Ans: True

2. Which of the following oxides is amphoteric?
   A) NO_2 B) Sb_2O_3 C) Bi_2O_3 D) As_4O_6 E) P_2O_5
   Ans: B

3. Beside each oxide, indicate whether it is acidic, basic, or amphoteric.
   P_2O_5
   NO_2
   Sb_2O_3
   As_4O_6
   Bi_2O_3
   Ans: P_2O_5 acidic
   NO_2 acidic
   Sb_2O_3 amphoteric
   As_4O_6 amphoteric
   Bi_2O_3 basic

4. Ammonia is both a weak Brønsted base in water and a strong Lewis acid.
   Ans: True

5. Which of the following is likely to dissolve to the greatest extent in liquid ammonia?
   A) KCl   B) $MgCl_2$ C) NaCl   D) RbI   E) LiF
   Ans: D

6. Which of the Group 15 elements are metalloids?
   A)   As only   D)   P, As, and Sb
   B)   As and Sb   E)   As, Sb, and Bi
   C)   P and As
   Ans: B

7. Autoprotolysis occurs to a much smaller extent in ammonia ($K_m = 1 \times 10^{-33}$ at $-35°C$) than in water. This means that
   A)   very strong bases that would be protonated in water survive in their anionic form in ammonia.
   B) $[NH_4^+(am)] \neq [NH_2^-(am)]$.
   C)   strongly ionic compounds are very soluble in ammonia.
   D)   ammonia is not very soluble in water.
   E)   ammonia is not a very good solvent for a wide range of compounds.
   Ans: A

8. Ammonium nitrate is explosive because
   A) the ammonium ion decomposes readily.
   B) the ammonium ion can be oxidized by the nitrate ion.
   C) the nitrate ion decomposes readily.
   D) the ammonium ion is readily reduced.
   E) nitrate reacts readily with oxygen in air.
   Ans: B

9. In the following reaction
   $$Cu^{2+}(aq) + 4NH_3(aq) \rightarrow Cu(NH_3)_4^{2+}(aq)$$
   the Lewis acid is _____ and the Lewis base is _____.
   Ans: The Lewis acid is Cu^{2+} and the Lewis base is ammonia.

10. Hydrazine is produced by the reaction of
    A) nitric acid with hygrogen.
    B) nitrogen dioxide with water.
    C) nitrogen with water vapor at high pressure.
    D) ammonia with alkaline hypochlorite solution.
    E) ammonia with hydrogen.
    Ans: D

11. The complex $(NH_3)_5RuN_2^{2+}$ can be synthesized by the reaction of hydrazine with $RuCl_3$. In this reaction, the reducing agent is _____ and the oxidizing agent is

    _____.

    Ans: The reducing agent is hydrazine and the oxidizing agent is $RuCl_3$.

12. Magnesium nitride dissolves in water to produce
    A) $HNO_3(aq)$ and $MgO(s)$.   D) $Mg(NO_3)_2(aq)$.
    B) $N_2(g)$ and $MgO(s)$.   E) $N_2H_4(l)$ and $Mg(OH)_2(s)$.
    C) $NH_3(g)$ and $Mg(OH)_2(s)$.
    Ans: C

13. The formula of hydrazine is
    A) N_2H_2.   B) HNO_2.   C) NH_2NH_2.   D) HN_3.   E) NH_2OH.
    Ans: C

14. Nitrogen triiodide can be prepared as the ammoniate, $NI_3 \bullet NH_3$. This compound is explosive when dry and decomposes according to the reaction below.

$$2NI_3 \bullet NH_3 \rightarrow N_2 + 2NH_3 + 3I_2$$

In this reaction, nitrogen undergoes a change in oxidation number from
A)   +3 to 0.
B)   +3 to −3.
C)   No change in oxidation number occurs because this is not a redox reaction.
D)   +6 to 0.
E)   +1 to 0.
Ans: A

15. The most stable form of phosphorus at 25°C and 1 atm is
A) P(s, red).   B) $P_4O_{10}(s)$.   C) $P_2O_3(s)$.   D) $P_4(g)$.   E) P(s, white).
Ans: E

16. Both nitrides and phosphides act as strong Brønsted bases when they react with water to form ammonia and phosphine, respectively.
Ans: True

17. Which of the following reactions are hydrolysis reactions?
    1. $PCl_3(l) + 3H_2O(l) \rightarrow H_3PO_3(aq) + 3 HCl(aq)$
    2. $Mg_3N_2(s) + 6H_2O(l) \rightarrow 3Mg(OH)_3(s) + 2NH_3(g)$
    3. $N_2H_4(aq) + O_2(g) \rightarrow N_2(g) + 2H_2O(l)$
    4. $SiF_4(g) + 2H_2O(l) \rightarrow SiO_2(s) + 4HF(aq)$
A) (1) only   B) (2) and (4)   C) (2) and (3)   D) (1) and (4)   E) (2) only
Ans: D

18. White phosphorus
    A)   is an amorphous substance.
    B)   is not very toxic.
    C)   is very unreactive.
    D)   consists of large, random aggregates of phosphorus atoms.
    E)   consists of tetrahedral P_4 molecules.
    Ans: E

19. Aqueous solutions of phosphine, PH_3,
    A)   are neutral.   D)   readily form $PH_4^+(aq)$.
    B)   contain the $PH_2^-(aq)$ ion exclusively.   E)   cannot exist.
    C)   are basic.
    Ans: A

20. Phosphorus pentachloride reacts with water to give
    A) $H_3PO_3(s)$ and $HCl(g)$.
    B) $H_3PO_4(l)$ and $HClO(aq)$.
    C) $H_3PO_3(s)$ and $Cl_2(g)$.
    D) $H_3PO_4(l)$ and $Cl_2(g)$.
    E) $H_3PO_4(l)$ and $HCl(g)$.
    Ans: E

21. The geometry of PCl_4^+ is
    A) T-shaped.
    B) tetrahedral.
    C) seesaw.
    D) trigonal bipyramidal.
    E) trigonal pyramidal.
    Ans: B

22. In the Ostwald process, the total change in oxidation number of nitrogen from reactant to nitric acid is
    A) 5.   B) 3.   C) 6.   D) 8.   E) 7.
    Ans: D

23. The first reaction in the Ostwald process for the synthesis of nitric acid is the
    A) oxidation of ammonia to nitrogen dioxide.
    B) reduction of ammonia with copper.
    C) decomposition of ammonia to nitrogen and hydrogen.
    D) production of $NH_3(l)$.
    E) oxidation of ammonia to nitrogen oxide.
    Ans: E

24. Dinitrogen trioxide can be prepared by
    A) reacting hydrazine with water.
    B) reacting ammonia with water.
    C) dehydrating nitrous acid.
    D) dehydrating nitric acid.
    E) reacting N_2O with O_2.
    Ans: C

25. Given that $N_2O_3(g)$ dissociates to $NO(g)$ and $NO_2(g)$, how many lone pairs of electrons are in the Lewis structure of N_2O_3?
    A) 2   B) 6   C) 1   D) 4   E) 8
    Ans: E

26. Dinitrogen trioxide is the anhydride of
    A) nitrous acid.
    B) nitric acid.
    C) dinitrogen tetroxide.
    D) hydrazine.
    E) hydrazoic acid.
    Ans: A

27. The anhydride of phosphoric acid is produced by the reaction
    A) $H_3PO_4(aq) \rightarrow HPO_3(aq) + H_2O(l)$.
    B) $2H_3PO_4(l) \rightarrow H_4P_2O_7(l) + H_2O(l)$.
    C) $P_4(s) + 3O_2(g) \rightarrow P_4O_6(s)$.
    D) $P_4(s) + 5O_2(g) \rightarrow P_4O_{10}(s)$.
    E) $P_4(g) + 3OH^-(aq) + 3H_2O(l) \rightarrow 3H_2PO_2-(aq) + PH_3(g)$.
    Ans: D

28. Phosphorus(V) oxide reacts with water to produce
    A) H_3PO_2.   B) H_3PO_4.   C) HPO_3.   D) H_3PO_3.   E) $H_3PO_2^-$.
    Ans: B

29. The formula of phosphorus(V) oxide is
    A) P_2O_4.   B) P_4O_6.   C) P_4O_{10}.   D) PO_2.   E) P_2O_3.
    Ans: C

30. Phosphorous and phosphoric acids have
    A)   1 and 0 hydrogens bonded to the phosphorus atom, respectively.
    B)   1 hydrogen bonded to the phosphorus atom.
    C)   3 hydrogens bonded to oxygen atoms.
    D)   2 and 0 hydrogens bonded to the phosphorus atom.
    E)   2 hydrogens bonded to the phosphorus atom.
    Ans: A

31. Draw the structures of the phosphate, phosphite, and hypophosphite ions.
    Ans: PO_4^{3-}, $H-PO_3^{2-}$, and $H_2PO_2^-$

32. Which of the following is diamagnetic?
    A) O_2^{2-} B) NO_2 C) O_2 D) O_2^-
    Ans: A

33. The term *catenation* refers to
    A)   the ability of sulfur to bond four oxygen atoms.
    B)   the ability of sulfur to form chains of atoms.
    C)   the fact that sulfur can exist as monatomic $S(g)$.
    D)   the fact that ozone is an allotrope of oxygen.
    E)   the synthesis of sulfuric acid.
    Ans: B

34. When sulfur is produced by the Claus process, the starting material is
    A) $H_2SO_4(l)$.   B) $Na_2SO_3(s)$.   C) $SO_3(g)$.   D) $SO_2(g)$.   E) $H_2S(g)$.
    Ans: E

35. Which of the following pairs are allotropes?
    A) $S_8(s)$ and $H_2S(g)$ D) $C(s)$ and $CO_2(g)$
    B) $S(s)$ and $SO_2(g)$ E) $O_2(g)$ and $O_3(g)$
    C) $O_2(g)$ and $H_2O_2(l)$
    Ans: E

36. Both H_2O and H_2S can participate in hydrogen bonding.
    Ans: False

37. When hydrogen peroxide disproportionates it forms water and oxygen gas. True or False?
    Ans: True

38. When sulfur is produced by the Claus process, the final step is
    A) the reduction of sulfuric acid by carbon.
    B) the oxidation of $H_2S(g)$ by $SO_2(g)$.
    C) the oxidation of $H_2S(g)$ by ozone.
    D) the oxidation of $H_2S(g)$ by oxygen.
    E) the reduction of $SO_2(g)$ by hydrogen.
    Ans: B

39. Which of the following is paramagnetic?
    A) N_2 B) N_2O C) O_2^{2-} D) O_3 E) NO
    Ans: E

40. In the *contact* process for the production of sulfuric acid, sulfur is first burned in oxygen to produce $SO_2(g)$. The $SO_2(g)$ is then
    A) reduced to $H_2S(g)$.
    B) dissolved in water to form *oleum*.
    C) dissolved in water to form $H_2SO_4(aq)$.
    D) dissolved in water to form $H_2SO_3(aq)$.
    E) oxidized to $SO_3(g)$.
    Ans: E

41. The starting material for the production of sulfuric acid by the *contact* process is
    A) $H_2S(g)$.   B) $H_2S_2O_7(l)$.   C) $S(g)$.   D) $SO_3(g)$.   E) $SO_2(g)$.
    Ans: C

42. The *contact* process for the production of sulfuric acid is
$$S(g) + O_2(g) \rightarrow SO_2(g)$$
$$2SO_2(g) + O_2(g) \rightarrow 2SO_3(g)$$
$$SO_3(g) + H_2SO_4(l) \rightarrow H_2S_2O_7(l)$$
$$H_2S_2O_7(l) + H_2O(l) \rightarrow 2H_2SO_4(l)$$
How many moles of "new" sulfuric acid are produced from 150 mol S(g)?
A) 300 mol   B) 450 mol   C) 225 mol   D) 150 mol   E) 75 mol
Ans: D

43. Sulfur does not react directly with
   A)   chlorine.   D)   fluorine and chlorine.
   B)   chlorine and bromine.   E)   iodine.
   C)   fluorine.
Ans: E

44. Sulfur dioxide can act as both an oxidizing agent and a reducing agent. Explain.
Ans: In sulfur dioxide, the oxidation state, +4, is intermediate in sulfur's range from −2 to +6.

45. The stable forms of chlorine, bromine, and iodine at 25°C and 1 atm are
   A) $Cl_2(g)$, $Br_2(g)$, and $I_2(g)$.   D) $Cl_2(g)$, $Br(l)$, and $I(s)$.
   B) $Cl_2(g)$, $Br_2(l)$, and $I_2(s)$.   E) $Cl_2(g)$, $Br(l)$, and $I_2(g)$.
   C) $Cl_2(g)$, $Br_2(l)$, and $I_2(l)$.
Ans: B

46. Which one of the following statements is false?
   A)   Fluorine can be produced by oxidation of a fluoride ion with chlorine.
   B)   Bromine can be produced by oxidation of a bromide ion with chlorine.
   C)   Fluorine is produced by electrolysis of an anhydrous mixture of KF and HF.
   D)   The most stable form of fluorine at 25°C and 1 atm is $F_2(g)$.
Ans: A

47. The lattice enthalpies of ionic compounds of fluoride tend to be very high because
   A)   fluorine has a high electronegativity.
   B)   fluoride ion has an oxidation state of −1.
   C)   fluorine gas is very reactive.
   D)   the fluoride ion is small.
   E)   fluorine is a strong oxidant.
Ans: D

48. All of the following are characteristic of fluorine or its compounds except
    A)   fluorine has a high electronegativity.
    B)   fluorine brings about low oxidation numbers in other elements.
    C)   fluorine has a very small size.
    D)   fluorine does not have available d-orbitals for bonding.
    E)   fluorine has an oxidation number of −1.
    Ans: B

49. All of the following acids have pK$_a$'s smaller than 2.0 except
    A) HIO.   B) HBrO$_3$.   C) HClO$_3$.   D) HIO$_4$.   E) HClO$_4$.
    Ans: A

50. Order the following acids from weakest to strongest: HClO, HClO$_4$, HClO$_2$, and HClO$_3$.
    A)   HClO < HClO$_3$ < HClO$_4$ < HClO$_2$ D)   HClO$_4$ < HClO$_3$ < HClO$_2$ < HClO
    B)   HClO$_4$ < HClO < HClO$_2$ < HClO$_3$ E)   HClO < HClO$_2$ < HClO$_3$ < HClO$_4$
    C)   HClO < HClO$_4$ < HClO$_3$ < HClO$_2$
    Ans: E

51. Which of the following is unstable?
    A) HFO   B) HBrO   C) HClO   D) HI   E) HF
    Ans: A

52. When sulfuric acid and calcium fluoride are mixed, a product is
    A) F$_2$(g).   B) SO$_2$(g).   C) HF(g).   D) H$_2$S(g).   E) SF$_2$(g).
    Ans: C

53. When phosphoric acid and potassium bromide are mixed, a product is
    A) HBrO(g).   B) HBr(g).   C) PBr$_2$(g).   D) Br$_2$(g).   E) P$_4$(s).
    Ans: B

54. Which of the following is the weakest acid?
    A) HClO$_4$ B) HIO   C) HBrO$_4$ D) HClO   E) HBrO$_3$
    Ans: B

55. Which of the following has the largest bond enthalpy?
    A) HCl   B) All the bond enthalpies are the same.   C) HBr   D) HF   E) HI
    Ans: D

56. The oxoacids of Group 17 have the general formula HXO_n where X is a halogen and n = 1 to 4. Which of the following is true?
    A)   As the oxidation number of X increases, the strength of the acid decreases.
    B)   As the oxidation number of X increases, the oxidizing strength of the acid increases.
    C)   Only the oxoacids with oxidation number +1 are strong acids.
    D)   The oxoacids with oxidation number +1 are reducing agents.
    E)   The oxoacids with oxidation number +7 are weak oxidizing agents.
    Ans: B

57. What is the shape of BrF_5?
    A)   seesaw   D)   octahedral
    B)   square planar   E)   trigonal bipyramidal
    C)   square pyramidal
    Ans: C

58. In ClF_3, the number of regions of high electron concentration around the chlorine atom are
    A) 3   B) 4   C) 5   D) 6   E) 2
    Ans: C

59. What is the shape of ClF_3?
    A)   trigonal planar   D)   T-shaped
    B)   trigonal bipyramidal   E)   trigonal pyramidal
    C)   tetrahedral
    Ans: D

60. The phase diagram for helium indicates that
    A)   helium has three phases.
    B)   helium has four phases.
    C)   helium can be liquefied at ordinary pressures.
    D)   helium exists as He(g) and He(l) at about 2 K.
    E)   helium exists only as a gas.
    Ans: B

61. The phase diagram for helium
    A)   indicates that helium exists as He(g) and He(l) at about 2 K.
    B)   indicates that helium exists only as a gas.
    C)   has two triple points.
    D)   indicates that helium can be liquefied at ordinary pressures.
    E)   indicates that helium has three phases.
    Ans: C

62. Which of the following noble gases is least likely to react with fluorine?
    A) Ar   B) Rn   C) Xe   D) Kr   E) Ne
    Ans: E

63. Which of the following noble gases is likely to have the smallest ionization energy?
    A) Xe   B) Kr   C) He   D) Ne   E) Ar
    Ans: A

64. When XeF_4 acts as a fluorinating agent,
    A) F_2 is produced.
    B) XeF_4 disproportionates.
    C) XeF_3^+ is produced and is the reactive species.
    D)   Xe is reduced.
    E)   Xe is oxidized.
    Ans: D

65. The reaction of $XeF_4(s)$ with $Pt(s)$ produces
    A) $Xe(g)$ and $PtF_6(s)$.   D) $Xe(g)$ and $PtF_4(s)$.
    B) $XeF_2(s)$ and $PtF_2(s)$.   E) $XeF_2(s)$ and $PtF_4(s)$.
    C)   dimeric $XeF_2PtF_4(s)$.
    Ans: D

66. What is the shape of XeF_5^+?
    A)   trigonal bipyramidal   D)   square pyramidal
    B)   square planar   E)   seesaw
    C)   T-shaped
    Ans: D

67. Xenon can exhibit all of the following oxidation numbers in its compounds.
    A)   +2, +4, +6, +8   D)   +2, +4, +6, +7
    B)   0, +2, +3   E)   +2, +3, +4, +5, +6
    C)   +2, +4, +5
    Ans: A

68. The gaseous halides can all be synthesized by oxidation of their ions.
    Ans: False

69. Bromine can be obtained from brines by oxidation of Br^- by Cl_2.
    Ans: True

70. Which of the following has the strongest molar bond enthalpy?
    A) H-F   B) H-At   C) H-Cl   D) H-I   E) H-Br
    Ans: A

71. Which of the following reacts only as a reducing agent?
    A) P_4 B) NO_2- C) F_2 D) Cl^- E) S_8
    Ans: D

72. What is the shape of ClF_4^+?
    A)   seesaw   D)   tetrahedral
    B)   octahedral   E)   square planar
    C)   trigonal bipyramidal
    Ans: A

73. When magnesium burns in air, what are the main products?
    Ans: magnesium oxide and magnesium nitride

74. Consider the following reactions (unbalanced).
    $$Na(s) + P_4(s) \rightarrow A$$
    $$A + H_2O(l) \rightarrow B + C$$
    Write the formulas of A, B, and C.
    Ans: A = $Na_3P(s)$, B = $PH_3(g)$, C = $NaOH(aq)$

75. (a) The phase diagram of helium has _____ triple points.
    (b) The ionization energy of both oxygen and xenon is about 1200 kJ·mol^{-1}. What does this observation suggest?
    (c) The first Noble gas compound actually contained the XeF^+ ion. Write the Lewis structure of this ion and predict whether it should be stable.
    Ans: (a) 2; (b) if O_2 can be oxidized in a reaction, then xenon should also react in the same reaction; (c) the Lewis structure is the same as that of F_2 and thus XeF^+ should be stable.

76. Ammonia is a weak Brønsted base but a strong Lewis base. True or False?
    Ans: True

77. Both ammonia and phosphine are soluble in water. Which is least soluble and why?
    Ans: PH_3 is least soluble because it cannot form hydrogen bonds with water.

78. Nitric acid can never act as a reducing agent. True or False?
    Ans: True

79. Which of the following is the anhydride of sulfurous acid?
    A) SO_2 B) SO_3 C) SO_4^{2-} D) HSO_4^-
    Ans: A

80. All of the following form clathrates with water except
    A) CH_4 B) SO_2 C) NO D) CO_2
    Ans: C

81. Which of the following is/are properties of sulfuric acid?
    A)   Sulfuric acid is an oxidizing agent.
    B)   Sulfuric acid is a strong dehydrating agent.
    C)   Sulfuric acid is a strong Bronsted acid.
    D)   B, C, and D are all properties of sulfuric acid.
    Ans: D

82. Because the sulfur atom in SF_6 has an oxidation number of +6, SF_6 is a good oxidizing agent. True or False?
    Ans: False

83. The preparation of HI(g) by the reaction of KI(s) with phosphoric acid is not a redox reaction. True or False?
    Ans: True

84. The products of the disproportionation of chlorine in water are
    A) HClO and O_2 B) HClO and HCl   C) ClO_2 and H_2 D) OH^- and HCl
    Ans: B

85. What was the key idea that lead to the synthesis of fluorine and oxygen compounds of xenon?
    Ans: The ionization energy of Xe is similar to that of F_2 and O_2.

Chapter 16: The Elements: The *d* Block

1. Most *d*-block elements have more than one common oxidation state.
   Ans: True

2. Except for scandium and titanium, all the 3d transition metals form divalent ions in aqueous solution.
   Ans: True

3. Which of the following is expected to be the best oxidizing agent?
   A) $Cu^+(aq)$ B) $Ti^{3+}(aq)$ C) $CrO_4^{2-}(aq)$ D) VO_2^+ E) $Cr^{2+}(aq)$
   Ans: C

4. For a given metal, the oxides become more basic with increasing oxidation state.
   Ans: False

5. Which of the following would be the strongest acid in aqueous solution?
   A) $Mg(OH_2)_6^{2+}(aq)$ D) $Sc(OH_2)_6^{3+}(aq)$
   B) $V(OH_2)_6^{2+}(aq)$ E) $Fe(OH_2)_6^{2+}(aq)$
   C) $Ni(OH_2)_6^{2+}(aq)$
   Ans: D

6. The metalloproteins hemoglobin and vitamin B_{12} contain the metals
   A)   iron and zinc, respectively.   D)   manganese and iron, respectively.
   B)   cobalt and nickel, respectively.   E)   iron and cobalt, respectively.
   C)   iron and Mo, respectively.
   Ans: E

7. The *d*-metals can be mixed together to form a wide range of alloys because
   A)   the *d*-electrons interact strongly with each other.
   B)   the *d*-metals have a wide range of metal radii.
   C)   the nucleus is well shielded by the d electrons.
   D)   the range of d metal radii is not very great.
   E)   the *d*-metals have low melting points.
   Ans: D

8. The atomic radii of the Period 6 *d*-metals are
   A)   smaller than expected, and about the same as the Period 5 *d*-metals.
   B)   about the same as the lanthanides.
   C)   larger than the atomic radii of the Period 5 *d*-metals.
   D)   about the same as the Period 4 *d*-metals.
   E)   large because of the lanthanide contraction.
   Ans: A

9. In the *d*-block elements, the third-row metallic radii are about the same as the second-row radii because of
   A) the lanthanide contraction.
   B) The third-row metallic radii are greater than the second-row radii.
   C) the increased number of *d*-electrons.
   D) greater shielding of *f*-electrons.
   E) greater shielding of *d*-electrons.
   Ans: A

10. The lanthanide contraction results because
    A) the shielding effect of the *d*-electrons is less than that of the *f*-electrons.
    B) the shielding effect of the *d*-electrons is greater than that of the *f*-electrons.
    C) of the poor shielding effect of the *f*-electrons and the higher nuclear charge.
    D) the *d*-electrons shield the *f*-electrons from the nucleus.
    E) *f*-electrons interact more strongly with each other than *d*-electrons.
    Ans: C

11. Which of the following has the smallest atomic radius?
    A) Ru   B) Pd   C) Mo   D) Cd   E) Zn
    Ans: E

12. The properties of the transition metals vary greatly but they all have
    A) high densities and high melting points.
    B) the same physical properties.
    C) oxides that are used to produce alloys.
    D) low boiling points.
    E) the same strength as their alloys.
    Ans: A

13. Which of the following pairs has about the same atomic radius?
    A) Cu and Ag   D) Mn and Tc
    B) Co and Rh   E) Ru and Os
    C) Fe and Ru
    Ans: E

14. Which of the following is likely to be a good oxidizing agent?
    A) Fe^{2+} B) Cr^{2+} C) V^{2+} D) Mn^{2+} E) MnO_4^-
    Ans: E

15. Which of the following is likely to be a good reducing agent?
    A) $Cr_2O_7^{2-}$ B) MnO_4^- C) Cr^{2+} D) MnO_2 E) Cr_2O_3
    Ans: C

16. The maximum oxidation states for chromium and manganese are
    A)   both +7.   D)   +5 and +6, respectively.
    B)   +7 and +8, respectively.   E)   +6 and +7, respectively.
    C)   +3 and +4, respectively.
    Ans: E

17. For the oxides CrO, Cr_2O_3, and CrO_3, which of the following is true?
    A)   All the oxides are basic.
    B)   All the oxides are acidic.
    C) CrO is acidic, and Cr_2O_3 and CrO_3 are basic.
    D) CrO and Cr_2O_3 are acidic and CrO_3 is basic.
    E) CrO is basic, Cr_2O_3 is amphoteric, and CrO_3 is acidic.
    Ans: E

18. Which of the following oxides is likely to be acidic?
    A) Mn_2O_7 B) MnO_2 C) Mn_2O_7 and MnO D) Mn_2O_3 E) MnO
    Ans: A

19. Vanadium metal is produced by the
    A)   reduction of $VOCl_2(s)$ with magnesium.
    B)   reduction of vanadium(V) oxide with calcium.
    C)   reaction of $V_2O_5(s)$ with coke.
    D)   decomposition of $VCl_4(g)$ at high temperature.
    E)   roasting of $VO_2(s)$.
    Ans: B

20. What is the formula of the compound formed when potassium dichromate is dissolved in aqueous base?
    A) K_2CrO_4 B) $K_2Cr_2O_7$ C) CrO_3 D) $Cr(OH)_3$
    Ans: A

21. Both $Cu^{2+}(aq)$ and $Zn^{2+}(aq)$ are colorless.
    Ans: True

22. The formulas of mercury(I) and mercury(II) chloride are
    A) Hg_2Cl_3 and $HgCl_2$, respectively.   D) Hg_2Cl and $HgCl_2$, respectively.
    B) Hg_2Cl_2 and $HgCl_2$, respectively.   E) Hg_2Cl_2 and $HgCl$, respectively.
    C) $HgCl$ and $HgCl_2$, respectively.
    Ans: B

23. In the smelting of copper, sulfur dioxide is removed by the following reaction.
    $$CaO(s) + SO_2(g) \rightarrow CaSO_3(s)$$
    In this reaction, the Lewis acid is _____ and the Lewis base is _____.
    Ans: $SO_2(g)$ and $CaO(s)$.

24. Zinc is used to galvanize iron. Explain why a galvanized container should not be used to store sodium hydroxide.
    Ans: Zinc is amphoteric and reacts with base to produce $Zn(OH)_4^{2-}$(aq).

25. A ligand is
    A) a Lewis base.
    B) necessary in a complex to balance the charge on the metal ion.
    C) used to force octahedral geometry on a metal ion.
    D) a Lewis acid.
    E) used to precipitate metal ions from aqueous solution.
    Ans: A

26. In the complex $[Fe(CN)_6]^{4-}$,
    A) iron is a Lewis acid and cyanide is a Lewis base.
    B) cyanide is a Lewis acid.
    C) iron has an oxidation number of +3.
    D) cyanide has an oxidation number of –2.
    Ans: A

27. All of the following ligands are monodentate except
    A) ammine.   B) cyano.   C) aqua.   D) carbonyl.   E) oxalato.
    Ans: E

28. What is the oxidation number of cobalt in $[CoCl(NH_3)_5]Cl_2$?
    A) +4   B) +1   C) +6   D) +2   E) +3
    Ans: E

29. What is the oxidation number of iron in $K_3[Fe(CN)_6]$?
    A) +3   B) +2   C) +9   D) +6
    Ans: A

30. What is the name of the complex $[Cr(en)_2(OH_2)Cl]Cl_2$?
    A) diethylenediaminebis(aquachloro)chromium(III) chloride
    B) aquachlorobis(ethylenediamine)chromium(III) chloride
    C) chloroaquabis(ethylenediamine)chromate(III) chloride
    D) aquachlorobis(ethylenediamine)chromium(II) dichloride
    E) aquachloridebis(ethylenediamine)chromium(III) chloride
    Ans: B

31. What is the name of the complex $Na[Co(NH_3)_3Cl_3]$?
    A)   sodium triamminetrichlorocobaltate(II)
    B)   sodium triamminetrichlorocobalt(III)
    C)   sodium tris(aminechloro)cobaltate(II)
    D)   sodium triamminetrichlorocobalt(II)
    E)   sodium trichlorotriamminecobalt(III)
    Ans: A

32. Which of the following is the formula of potassium hexacyanoferrate(II)?
    A) $[Fe(CN)_6]$ D) $K_3[Fe(CN)_6]$
    B) $K[Fe(CN)_6]$ E) $K_4[Fe(CN)_6]$
    C) $K_2[Fe(CN)_6]$
    Ans: E

33. One mole of *trans*-tetraamminedichlorocobalt(III) chloride dissolves in water to give
    A)   2 moles of ions.   C)   1 moles of ions.
    B)   3 moles of ions.   D)   4 moles of ions.
    Ans: A

34. One mole of pentaamminechlorocobalt(III) chloride dissolves in water to give _____ moles of ions.
    Ans: 3

35. Complexes with coordination number 4 are either tetrahedral or octahedral.
    Ans: False

36. Which of the following are ionization isomers?
    A)   cis-$[CoCl_2(en)_2]Cl$ and trans-$[CoCl_2(en)_2]Cl$
    B) $[Cr(NH_3)_6][Fe(CN)_6]$ and $[Fe(NH_3)_6][Cr(CN)_6]$
    C) $[CrCl(OH_2)_5]Cl_2 \cdot H_2O$ and $[CrCl_2(OH_2)_4]Cl \cdot 2H_2O$
    D) $[CoBr(NH_3)_5]SO_4$ and $[CoSO_4(NH_3)_5]Br$
    E) $[CoNCS(NH_3)_5]Cl_2$ and $[CoSCN(NH_3)_5]Cl_2$
    Ans: D

37. How many geometric isomers are possible for the complex ion $[Co(NH_3)_3(OH_2)_3]^{3+}$?
    A) 4   B) 1   C) 0   D) 2   E) 3
    Ans: D

38. How many geometric isomers are possible for the complex $[Co(en)(OH_2)Cl_3]$?
    A) 3   B) 2   C) 4   D) 5   E) 0
    Ans: B

39. The ligand 4-cyanopyridine can form _____ complex(es) with penataamminecobalt(III).
    Ans: 2

40. How many different isomers of all types are possible for the complex ion
    $[Co(NCS)_2(NH_3)_4]^+$?
    A) 6   B) 9   C) 4   D) 3   E) 2
    Ans: A

41. Consider the complexes *fac*- and *mer*-triaquatrichlororuthenium(III) chloride. Neither of these complexes is chiral.
    Ans: True

42. How many geometric isomers are possible for tris(glycinato)cobalt(III)?
    A) 5   B) 2   C) 0   D) 3   E) 4
    Ans: B

43. There are two geometric isomers of tris(glycinato)cobalt(III). Neither of these isomers is chiral.
    Ans: True

44. Which of the following complexes is chiral?

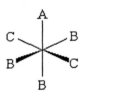 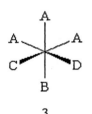 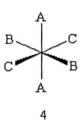

    A) 4   B) 3   C) 1 and 3   D) 2   E) 1 and 2
    Ans: B

45. Which of the following complexes is chiral

 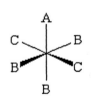 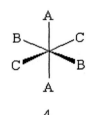

    A) 1   B) 4   C) None of the complexes is chiral.   D) 2   E) 2 and 3
    Ans: C

46. Which of the following complexes is chiral?

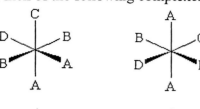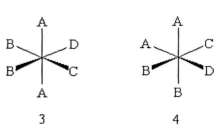

1   2   3   4

A) 4   B) 1 and 3   C) 2   D) 3   E) 1

Ans: A

47. For the complex ion $[Co(en)(NH_3)_2(OH_2)Cl]^{2+}$, there are four possible geometric isomers. How many of these geometric isomers are optically active?
A) none   B) 3   C) all of them   D) 1   E) 2

Ans: E

48. Which of the following complexes is chiral?
A)   trans-$[Cr(en)_2Cl_2]^+$ D)   trans-$[Cr(en)_2(NH_3)Cl]^{2+}$
B)   cis-$[Cr(en)_2Cl_2]^+$ E) $[Co(OH_2)_3(NH_3)_3]^{3+}$
C)   cis-$[Cr(NH_3)_4Cl_2]^+$

Ans: B

49. Which of the following complexes is chiral?
A) $[Co(en)_3]^{3+}$ B) $[RuCl_3(CN)_3]^{3-}$ C) $[Co(NH_3)_6]^{3+}$ D) trans-$[Cr(en)_3]^{3+}$

Ans: A

50. The complex $[Ru(OH_2)_3Cl_3]$
A)   is chiral.
B)   has 2 geometric isomers, neither of which is chiral.
C)   has 2 geometric isomers, one of which is chiral.
D)   has 2 geometric isomers, both of which are chiral.
E)   has no geometric isomers.

Ans: B

51. Which of the following complex ions has the largest *d*-orbital splitting energy?
A) $[Co(NH_3)_6]^{3+}$ D) $[CoCl_6]^{3-}$
B) $[Co(CN)_6]^{3-}$ E) $[Co(OH_2)_6]^{3+}$
C) $[Co(ox)_3]^{3-}$

Ans: B

52. Which of the following complex ions has the smallest *d*-orbital splitting energy?
A) $[CoCl_4]^{2-}$ D) $[Co(OH_2)_6]^{2+}$
B) $[CoCl_6]^{4-}$ E) $[Co(CN)_6]^{4-}$
C) $[Co(NH_3)_6]^{2+}$

Ans: A

53. Which of the following complexes has the smallest *d*-orbital splitting energy?
    A) $Co(NH_3)_6{}^{3+}$
    B) $Co(NH_3)_5OH^{2+}$
    C) $Co(NH_3)_5CN^{2+}$
    D) $Co(NH_3)_5I^{2+}$
    E) $Co(NH_3)_5(OH_2)^{3+}$
    Ans: D

54. Which of the following complex ions is likely to absorb at 600–700 nm?
    A) $[Co(OH_2)_6]^{2+}$
    B) $[Co(NO_2)_6]^{4-}$
    C) $[Co(NH_3)_6]^{2+}$
    D) $[Co(CN)_6]^{4-}$
    E) $[CoCl_4]^{2-}$
    Ans: E

55. Which of the following complex ions is likely to absorb at about 400 nm?
    A) $[Cr(OH_2)_6]^{3+}$ B) $[Cr(CN)_6]^{3-}$ C) $[CrF_6]^{3-}$ D) $[CrCl_6]^{3-}$ E) $[Cr(ox)_3]^{3-}$
    Ans: B

56. Which of the following complex ions is colorless?
    A) $[Cr(OH_2)_6]^{3+}$
    B) $[Ti(OH_2)_6]^{3+}$
    C) $[Ni(OH_2)_6]^{2+}$
    D) $[V(OH_2)_6]^{2+}$
    E) $[Zn(OH_2)_6]^{2+}$
    Ans: E

57. Consider the data below.

Complex	Absorbance maximum
$Ti(OH_2)_6{}^{3+}$	510 nm
$Fe(OH_2)_6{}^{3+}$	700 nm
$Fe(CN)_6{}^{4-}$	305 nm

    Which complex has the largest ligand field splitting?
    Ans: $Fe(CN)_6{}^{4-}$

58. For which one of the following would it not be possible to distinguish between high-spin and low-spin complexes in octahedral geometry?
    A) Cr(II) B) Co(III) C) Co(II) D) Fe(II) E) Ni(II)
    Ans: E

59. How many unpaired electrons are predicted for $[Co(OH_2)_6]^{2+}$?
    A) 0 B) 5 C) 2 D) 1 E) 3
    Ans: E

60. How many unpaired electrons are predicted for $[Mn(CN)_6]^{4-}$?
    A) 1 B) 2 C) 3 D) 5 E) 0
    Ans: A

61. A complex that is the subject of a great deal of research because of its high oxidation potential is ferrate, FeO_4^{2-}. How many unpaired electrons are in this complex?
    A) 4   B) 2   C) 1   D) 6   E) 3
    Ans: B

62. If an iron(II) complex is tetrahedral, how many unpaired electrons are predicted?
    A)   1
    B)   five or one, depending on the ligands
    C)   5
    D)   zero or four, depending on the ligands
    E)   4
    Ans: E

63. The complex ion $[Cr(OH_2)_6]^{2+}$ has four unpaired electrons. This means that
    A)   the complex is high-spin.
    B)   the complex is diamagnetic.
    C)   the water ligands are difficult to remove.
    D)   the complex is low-spin.
    E) Δ_o is very large.
    Ans: A

64. Which of the following is diamagnetic?
    A) $[Fe(OH_2)_6]^{2+}$ D) $[Co(NH_3)_6]^{3+}$
    B) $[Fe(CN)_6]^{3-}$ E) $[Cr(NH_3)_6]^{3+}$
    C) $[Co(OH_2)_6]^{2+}$
    Ans: D

65. Rhodium lies below cobalt in the periodic table. What is the *d*-electron configuration of $[Rh(CN)_6]^{3-}$?
    A) $t_{2g}^5 e_g^2$ B) $t_{2g}^4 e_g^1$ C) t_{2g}^6 D) $t_{2g}^4 e_g^2$ E) t_{2g}^5
    Ans: C

66. What is the *d*-electron configuration of the tetrahedral complex ion $[FeCl_4]^-$?
    A) $e^2 t^3$ B) $e^4 t^1$ C) e^5 D) $e^1 t^4$ E) $e^3 t^2$
    Ans: A

67. Comparing $[Co(CN)_6]^{3-}$ with $[CoCl_6]^{4-}$, which of the following statements is true?
    A) $[Co(CN)_6]^{3-}$ has more *d*-electrons than $[CoCl_6]^{4-}$.
    B) $[Co(CN)_6]^{3-}$ is paramagnetic while $[CoCl_6]^{4-}$ is diamagnetic.
    C)   Both complexes are paramagnetic.
    D) $[Co(CN)_6]^{3-}$ is diamagnetic while $[CoCl_6]^{4-}$ is paramagnetic.
    E) $[Co(CN)_6]^{3-}$ has the same number of *d*-electrons as $[CoCl_6]^{4-}$.
    Ans: D

68. All of the following complexes are high-spin except
    A) $V(OH_2)_6^{3+}$.
    B) $Zn(OH_2)_6^{2+}$.
    C) $Ni(OH_2)_6^{2+}$.
    D) $Co(OH_2)_6^{2+}$.
    E) $Fe(OH_2)_6^{3+}$.
    Ans: B

69. All of the following complexes are low-spin except
    A) $Co(CN)_6^{3-}$.
    B) $Co(NH_3)_6^{3+}$.
    C) $Rh(NH_3)_6^{3+}$.
    D) $Co(en)_3^{3+}$.
    E) $CoCl_4^{2-}$.
    Ans: E

70. How many unpaired electrons are expected for $Co(OH_2)_6^{2+}$?
    A) 0   B) 2   C) 4   D) 1   E) 3
    Ans: E

71. The complex $CoCl_4^{2-}$ absorbs light at approximately _____ nm and its color is

    _____.

    Ans: in the range of 600–700; violet to blue.

72. If $AgNO_3(aq)$ is added to a 1 M aqueous solution of pentaamminechlorocobalt(III) chloride, how many moles of silver chloride are precipitated?
    Ans: 2

73. The ligand CN^- is a strong-field ligand because it is a _____ and Cl^- is a weak-field ligand because it is a _____.
    Ans: π-acceptor; σ-donor

74. Consider the complex *trans*-$[Co(NH_3)_4Cl_2]Cl$. Draw the structure of the complex and clearly label or give the following.
    a) the inner and outer coordination spheres
    b) the coordination number
    c) the *trans* ligands
    d) How many stereoisomers are there for a complex with the formula $[Co(NH_3)_4Cl_2]Cl$?
    Ans: (a) the inner coordination sphere is everything inside the square brackets; (b) 6;
         (c) Cl^-; (d) 2; 2 geometric and 1 enantiomer

75. The compound Co(en)$_2$(NO$_2$)$_2$Cl has been prepared in a number of isomeric forms. All of the forms have the nitrite ion coordinated trough the nitrogen atom. Draw the structure of each isomeric form.
    a) One form is optically inactive and does not react with AgNO$_3$ or en.
    b) A second form reacts with AgNO$_3$ but not with en and is also optically inactive.
    c) A third form is optically active and reacts with both AgNO$_3$ and en.
    Ans: (a) *trans*-chloronitrobis(ethylenediamine)cobalt(III) nitrite; (b) *trans*-dinitrobis(ethylenediamine)cobalt(III) chloride; (c) *cis*-dinitrobis(ethylenediamine)cobalt(III) chloride

76. All rhodium complexes are low spin. True or False?
    Ans: True

77. How many unpaired electrons are there in [CoBr$_4$]$^{2-}$?
    A) 0   B) 2   C) 3   D) 4
    Ans: C

78. Give two reasons why Δ_T is always less than Δ_O.
    Ans: 1) There are two fewer ligands surrounding the metal. 2) The d orbitals do not point directly at the ligands.)

79. Which complex, [CoBr$_4$]$^{2-}$ or [Co(NH$_3$)$_5$]$^{3+}$, would be the most highly colored?
    Ans: [CoBr$_4$]$^{2-}$

80. A π-bonding ligand always gives a larger ligand field splitting energy than a σ-bonding ligand. True or False?
    Ans: False

81. The ligand field splitting energy is greater for complexes of π-acid ligands than for complexes of π-base ligands. True or False?
    Ans: True

82. Which of the following complexes is colorless?
    A) Fe(OH$_2$)$_6$$^{2+}$ B) Cu(OH$_2$)$_6$$^{+}$ C) Cu(OH$_2$)$_6$$^{2+}$ D) Ti(OH$_2$)$_6$$^{3+}$
    Ans: B

83. Which of the following complexes is likely to have an absorbance maximum at 305 nm?
    A) Fe(CN)$_6$$^{4-}$ B) Ti(OH$_2$)$_6$$^{3+}$ C) Fe(OH$_2$)$_6$$^{3+}$ D) Co(NH$_3$)$_6$$^{3+}$
    Ans: A

84. Which of the following complexes is high-spin?
    A) Cu(en)$_3$$^{2+}$ B) Cr(OH$_2$)$_6$$^{3+}$ C) Ni(en)$_3$$^{2+}$ D) Cr(OH$_2$)$_6$$^{2+}$
    Ans: D

85. Match each complex with the number of unpaired electrons.

$NiCl_4^{2-}$	0
$Fe(CN)_6^{3-}$	1
$Co(en)_3^{3+}$	2
$FeBr_4^{2-}$	4

Ans:

$NiCl_4^{2-}$	2
$Fe(CN)_6^{3-}$	1
$Co(en)_3^{3+}$	0
$FeBr_4^{2-}$	4

Chapter 17: Nuclear Chemistry

1. Heavy elements are more likely to give off β radiation.
   Ans: False

2. Heavy elements are more likely to give off α radiation.
   Ans: True

3. When a positron encounters an electron, both are annihilated and converted to energy.
   Ans: True

4. How many nucleons does boron-11 contain?
   A) 6   B) 10   C) 5   D) 11   E) 16
   Ans: D

5. How many nucleons does an alpha particle contain?
   A) 4   B) 0   C) 2   D) 8   E) 6
   Ans: A

6. An α particle corresponds to
   A) $_2^4\text{He}$.   B) $_1^1\text{H}^+$.   C) $_1^0\text{e}$.   D) $_{-1}^0\text{e}$.   E) $_2^4\text{He}^{2+}$.
   Ans: E

7. A β particle corresponds to
   A) $_{-1}^0\text{e}$.   B) $_1^0\text{H}^+$.   C) $_1^0\text{e}$.   D) $_0^1\text{n}$.   E) $_2^4\text{He}^{2+}$.
   Ans: A

8. When an α particle is emitted, the mass number
   A)   decreases by 4.   D)   decreases by 1.
   B)   increases by 2.   E)   increases by 4.
   C)   decreases by 2.
   Ans: A

9. When an β particle is emitted, the atomic number
   A)   increases by 4.   D)   decreases by 4.
   B)   decreases by 1.   E)   increases by 2.
   C)   decreases by 2.
   Ans: C

10. What nuclide undergoes α decay to produce the daughter nuclide radon-222 (Z = 86)?
    A) radium-226   B) thorium-230   C) polonium-222   D) protactium-230
    Ans: A

11. When ^{131}I emits a β particle, what nuclide is produced?
    A) ^{130}I   B) ^{127}Sb   C) ^{131}Xe   D) ^{131}Te   E) ^{130}Te
    Ans: C

12. The nuclear equation for the disintegration of U-238 produces Th-234. What other nuclide is produced?
    A) ^{4}He   B) neutron   C) positron   D) ^{3}He   E) β
    Ans: A

13. If U-238 undergoes α emission, what other nuclide is produced?
    A) Np-236   B) Ac-236   C) Th-234   D) U-234   E) Pa-234
    Ans: C

14. The nuclear equation for the disintegration of In-116 produces Sn-116 and
    A)   a positron.   D)   gamma rays.
    B)   helium-4.   E)   an electron.
    C)   a neutron.
    Ans: E

15. A nuclide undergoes α decay and forms ^{110}I. What is the nuclide?
    A) ^{112}Cs   B) ^{114}Cs   C) ^{110}Te   D) ^{114}I   E) ^{110}Xe
    Ans: B

16. A nuclide undergoes positron emission to form ^{22}Ne. What is the nuclide?
    A) ^{23}Mg   B) ^{18}O   C) ^{22}Na   D) ^{22}F   E) ^{23}Na
    Ans: C

17. A nuclide undergoes proton emission to form ^{52}Fe. What is the nuclide?
    A) ^{52}Mn   B) ^{53}Fe   C) ^{56}Ni   D) ^{52}Co   E) ^{53}Co
    Ans: E

18. What nuclide is formed when ^{90}Sr undergoes β decay?
    A) ^{86}Kr   B) ^{90}Y   C) ^{94}Zr   D) ^{89}Sr   E) ^{90}Rb
    Ans: B

19. Bombarding ^{54}Fe with a neutron results in emission of a proton and formation of
    A) ^{55}Fe.   B) ^{54}Cr.   C) ^{54}Co.   D) ^{54}Mn.   E) ^{49}Ti.
    Ans: D

20. What nuclide is formed when ^{234}Th undergoes β decay?
    A) ^{232}Pu   B) ^{234}Pa   C) ^{234}U   D) ^{233}Th   E) ^{230}Rn
    Ans: B

21. What nuclide is formed when ^{211}Po undergoes α decay?
    A) ^{207}Pb   B) ^{206}Ra   C) ^{206}Pb   D) ^{208}Rn
    Ans: A

22. What type of particle is emitted in the transformation below?
    $$^{201}Pt \rightarrow {}^{201}Au$$
    A) γ particle
    B)   positron
    C)   No particle is emitted because electron capture occurs.
    D) β particle
    E) α particle
    Ans: D

23. What type of particle is captured and what type is emitted in the nucleosynthesis below?
    $$Fe \rightarrow {}^{54}Mn$$
    A)   Neutron capture and proton emission.
    B)   Neutron capture and β particle emission.
    C)   Only neutron capture occurs.
    D)   Neutron capture and positron emission occurs.
    E)   Electron capture and β particle emission occurs.
    Ans: A

24. All of the following are predicted to be unstable except
    A) Ca-40   B) N-16   C) Na-24   D) O-15
    Ans: A

25. Which of the following is expected to be stable?
    A) ^{208}Pb   B) ^{7}Be   C) ^{122}Sb   D) ^{87}Kr   E) ^{24}Na
    Ans: A

26. Which of the following is most likely to undergo β decay?
    A) ^{222}Rn   B) ^{11}Li   C) ^{235}U   D) ^{114}Cs   E) ^{226}Ra
    Ans: B

27. Which of the following is most likely to undergo α decay?
    A) ^{24}Na   B) ^{47}K   C) ^{14}C   D) ^{232}Th   E) ^{3}H
    Ans: D

28. Elements with an even number of protons and neutrons
    A)   are usually stable.
    B)   are usually unstable.
    C)   never undergo α emission.
    D)   always have a proton to neutron ratio of about one.
    E)   have a magic number of protons or neutrons.
    Ans: A

29. Nuclides that lie below the *band of stability* decay by positron emission, electron capture, or proton emission. True or False?
    Ans: True

30. If a nuclide lies above the *band of stability*, it should decay by
    A)   They do not decay.   D)   neutron emission.
    B) β particle emission.   E) α particle emission.
    C)   positron emission.
    Ans: B

31. Which of the following nuclei is stable?
    A) ^{120}Sn B) ^{213}Fr C) ^{38}Cl D) ^{24}Na E) ^{15}O
    Ans: A

32. All of the following are stable except
    A) ^{15}O.   B) ^{20}Ne.   C) ^{14}N.   D) ^{12}C.   E) ^{16}O.
    Ans: A

33. What is the element produced when ^{83}Sr undergoes positron emission?
    A) Kr   B) Br   C) Rb   D) Zr   E) Y
    Ans: C

34. What is the element produced when ^{44}Ti undergoes electron capture?
    A) Ar   B) K   C) Sc   D) V   E) Ca
    Ans: C

35. Which of the following is most penetrating?
    A) α B) β C) β^+ D) p   E) n
    Ans: E

36. What process occurs when ^{222}Rn forms the daughter nuclide ^{218}Po?
    A)   electron capture   D)   neutron emission
    B) β emission   E) α emission
    C)   positron emission
    Ans: E

37. What is the identity of the product, P, in the following reaction?
    $$^{58}Fe + 2\,^{1}_{0}n \rightarrow P + \,^{0}_{-1}e$$
    Ans: ^{60}Co

38. Boron neutron capture therapy is used to destroy tumors. In this reaction ^{10}B absorbs a slow neutron to produce $^{11}B^*$ which produces ^{7}Li and _____ plus high-energy photons.
    Ans: α particles

39. Boron neutron capture therapy is used to destroy tumors. In this reaction ^{10}B absorbs a slow neutron to produce $^{11}B^*$ which produces α particles and _____ plus high-energy photons.
    Ans: ^{7}Li

40. Technetium-99m, a "metastable" nuclide of Tc is produced by the reaction below.
    _____ → $^{99m}Tc + \beta$
    What is the starting nuclide?
    Ans: ^{99}Mo

41. The outcome of positron emission and electron capture is the same; i.e. the atomic number is reduced by one.
    Ans: True

42. The age of any object can be determined by carbon-14 dating.
    Ans: False

43. In the ^{238}U decay series, there are eight α particles and six β particles lost starting with α emission from ^{238}U. What is the final product?
    A) ^{207}Pb B) ^{210}Po C) ^{210}Bi D) ^{208}Pb E) ^{206}Pb
    Ans: E

44. When ^{214}Pb decays as part of the ^{238}U decay series, there are two α particles and four β particles lost starting with β emission from ^{214}Pb. What is the final product?
    A) ^{207}Pb B) ^{206}Pb C) ^{210}Bi D) ^{210}Po E) ^{210}Pb
    Ans: B

45. Cobalt-60, used in the radiation treatment of cancer, is produced by absorption of two neutrons and emission of a β particle. What is the starting nuclide?
    A) ^{58}Mn B) ^{58}Ni C) ^{58}Fe D) ^{59}Co E) ^{59}Fe
    Ans: C

46. Radioactive decay is a
    A)   first-order process.   D)   temperature-dependent process.
    B)   second-order process.   E)   zero-order process.
    C)   non-spontaneous process.
    Ans: A

47. A 9.9-g sample of iodine-131 is stored for exactly 3 weeks. If the decay constant is $0.0861 \ day^{-1}$, what mass of the isotope remains?
    A) 0.15 g   B) 7.6 g   C) 1.6 g   D) 2.6 g   E) 5.5 g
    Ans: C

48. The decay constant for strontium-90 is 0.0247 y^{-1}. How many grams of strontium-90 remain if 10.0 g decay for exactly 60 years?
    A) 0.23 g   B) 0.33 g   C) 1.48 g   D) 2.27 g   E) 7.71 g
    Ans: D

49. The half-life of strontium-90 is 28.1 years. Calculate the percent of a strontium sample left after 100 years.
    A) 76%   B) 0.34%   C) 63%   D) 82%   E) 8.5%
    Ans: E

50. The half-life of plutonium-239 is 24,100 years. Calculate the percent of a plutonium sample left after 100 years.
    A) 99.7%   B) 4.15%   C) 90.0%   D) 50.0%   E) 96.4%
    Ans: A

51. Calculate the time required for the activity of a 9.0-mCi cobalt-60 source to decay to 8.5-mCi. The half-life of cobalt-60 is 5.26 years.
    A)   4.6 months   D)   10 months
    B)   2.3 months   E)   0.090 months
    C)   5.2 months
    Ans: C

52. Calculate the time required for the activity of a 9.0-mCi sodium-25 source to decay to 7.0-mCi. The half-life of sodium-25 is 60.0 s.
    A) 22 s   B) 44 s   C) 0.029 s   D) 19 s   E) 9.4 s
    Ans: A

53. Use the law of radioactive decay to determine the activity of a 2.0-µg sample of Tc-99, $t_{\frac{1}{2}} = 2.2 \times 10^5$ y.
    A) 1.0 Ci   B) 1.0×10^{-6} Ci   C) 1.5 Ci   D) 3.3×10^{-8} Ci   E) 3.3×10^{-2} Ci
    Ans: D

54. A sample of a dish from an archaeological *dig* gave 18,500 disintegrations per gram of carbon per day. A 1.00-g sample of carbon from a modern source gave 22,080 disintegrations per day. If the decay constant of carbon-14 is 1.2×10^{-4} y^{-1}, what is the age of the dish?
    A) 1500 y   B) 8300 y   C) 68000 y   D) 640 y   E) 30000 y
    Ans: A

55. What is the time needed for the activity of a radium-226 source to change from 1.5 Ci to 0.50 Ci? The half-life of radium-226 is 1.60×10^3 y.
    A) 5.33×10^2 y   D) 9.36×10^2 y
    B) 1.60×10^3 y   E) 2.54×10^3 y
    C) 1.10×10^3 y
    Ans: E

56. The binding energy of a nucleus is
    A)   the energy evolved when a nucleus breaks up into fundamental particles.
    B)   the energy released when Z protons and A−Z neutrons come together to form a nucleus.
    C)   related to the number of electrons in a nucleus.
    D)   the energy required to eject one neutron from the nucleus.
    E)   related to the mass lost in a chemical reaction.
    Ans: B

57. The nuclear binding energy for lithium-7 is the energy released in the nuclear reaction
    A) $3\,^1H + 7n \rightarrow\,^7Li$ D) $3\,^1H + 4n \rightarrow\,^7Li$
    B) $^6Li + n \rightarrow\,^7Li$ E) $7\,^1H \rightarrow\,^7Li$
    C) $3\,^1H + 4\beta \rightarrow\,^7Li$
    Ans: D

58. The nuclear binding energy for iron-56 is the energy released in the nuclear reaction
    A) $26\,^1H + 30\beta \rightarrow\,^{56}Fe$ D) $26\,^1H + 56n \rightarrow\,^{56}Fe$
    B) $56\,^1H \rightarrow\,^{56}Fe$ E) $26\,^1H + 30n \rightarrow\,^{56}Fe$
    C) $^{55}Fe + n \rightarrow\,^{56}Fe$
    Ans: E

59. The nuclear binding energy for neon-20 is the energy released when
    A)   neon-19 and 1 neutron form neon-20.
    B)   10 protons and 10 neutrons form neon-20.
    C)   10 protons and 10 electrons form neon-20.
    D)   20 neutrons form neon-20.
    E)   20 protons form neon-20.
    Ans: B

60. Calculate the nuclear binding energy of 1 mol lithium-7.
    A) 6.31×10^{-12} kJ   D) 1.14×10^{18} kJ
    B) 1.14×10^{15} kJ   E) 3.80×10^9 kJ
    C) 6.31×10^{-15} kJ
    Ans: E

61. Calculate the energy change for the synthesis of 1 mol ^{60}Co:
    $$^{58}Fe + 2n \rightarrow\,^{60}Co + \beta$$
    A) 2.53×10^{-15} kJ
    B) 1.52×10^9 kJ
    C) 2.53×10^{-12} kJ
    D)   The calculation cannot be done without the mass of β.
    E) 8.43×10^{-24} kJ
    Ans: B

62. Calculate the energy change when one ^{235}U nucleus undergoes the fission reaction
$$^{235}U + n \rightarrow \,^{142}Ba + \,^{92}Kr + 2n$$
The masses needed are ^{235}U, 235.04 u; ^{142}Ba, 141.92 u; ^{92}Kr, 91.92 u; n, 1.0087 u.

A) -3.2×10^{-28} J   D) $+2.9 \times 10^{-11}$ J

B) $+1.8 \times 10^{-10}$ J   E) -2.8×10^{-11} J

C) -1.7×10^{-10} J

Ans: E

63. The equation
$$6\,D \rightarrow 2\,^{4}He + 2\,^{1}H + 2\,n$$
represents

A) the binding energy of He.   D) fission.

B) fusion.   E) the decay of a proton.

C) electron capture.

Ans: B

64. One Becquerel is equal to _____ and one Curie is equal to

_____.

Ans: 1 disintegration per second; 3.7×10^{10} disintegrations per second

65. Nuclides that lie below the band of stability are proton rich and can form more stable nuclei by 1) _____,

2) _____ and

3) _____.

Ans: (1) β^{+} emission, (2) electron capture, and (3) proton emission

66. Which of the following nuclei is likely to capture an electron?

A) ^{9}Li   B) ^{41}Ca   C) ^{246}Am   D) ^{80}Ge   E) ^{24}Mg

Ans: B

67. Which of the following nuclei is likely to emit a positron?

A) ^{246}Am   B) ^{41}Ca   C) ^{118}Sn   D) ^{60}Cu   E) ^{20}Ne

Ans: D

68. Complete the following nuclear reaction.
$$\underline{\hspace{2cm}} + \,^{2}H \rightarrow \,^{97}Tc + 2n$$

Ans: ^{97}Mo

69. A 1.00-g sample of carbon from a modern source gave 15.3 disintegrations per minute. A sample of carbon from an "old" source gave 920 disintegrations per hour. What is the age of the "old" sample of carbon? The half-life of carbon-14 is 5.73×10^{3} year.

Ans: This is a modern source.

70. A sample of carbon from the Lascaux cave in France contained 12% of the original fraction carbon-14. Estimate the age of this sample. The half-life of carbon-14 is 5.73×10^3 year.
    Ans: 17,500 years

71. For the fusion of two deuterium atoms to produce helium-4, the difference in mass between products and reactants is -0.0256 u. What is E per mole of helium?
    Ans: 2.30×10^9 kJ

72. For the following thermonuclear reaction
    $$D + T \rightarrow {}^4He + n$$
    the difference in mass between products and reactants is -0.0189 u. What is ΔE per mole of helium?
    Ans: 1.70×10^9 kJ

73. The radioactive output of 1.00 g of radium-226 is defined as the Curie. Calculate this value in disintegrations per second. The half-life of radium-226 is 1600 years.
    Ans: 3.7×10^{10} disintegrations per second

74. What type of particle is emitted in the reaction below? Give an example of this process.

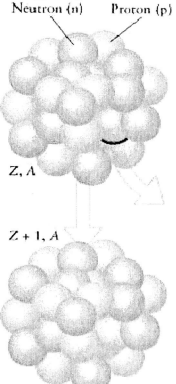

Ans: β particle

75. What type of particle is emitted in the reaction below? Give an example of this process.

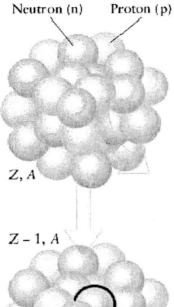

Neutron (n)   Proton (p)

Z, A

Z – 1, A

Ans: β^+ particle

76. What nuclide is formed when ^{24}Na undergoes β decay?
    A) ^{28}S   B) ^{24}Ne   C) ^{28}P   D) ^{24}Mg
    Ans: D

77. In the following equations, fill in the missing species.
    (a) ^{41}Ca + ___ $\rightarrow$ ^{41}K
    (b) ^{15}O $\rightarrow$ ^{15}N + ___
    Ans: (a) $^{0}_{-1}$e
         (b) $^{0}_{+1}$e

78. In the following equations, fill in the missing species.
    (a) ___ + $^{0}_{-1}$e $\rightarrow$ ^{7}Li
    (b) ^{9}Li $\rightarrow$ ___ + $^{0}_{-1}$e
    Ans: (a) ^{7}Be
         (b) ^{9}Be

79. An example of a particle that is referred to as being "doubly magic" is Pb-208. True or False?
    Ans: True

80. Why does the *band of stability* curve upward at high atomic number?
    Ans: Excess neutrons are required to overcome the mutual repulsion of the protons.

81. The nuclide ^{29}P decays by ejecting a β particle. True or False?
    Ans: False

82. The nuclide ^{7}Be decays by electron capture. True or False?
    Ans: True

83. Nuclides that lie above the *band of stability* are proton rich. True or False?
    Ans: False

84. Plutonium is one of the most toxic radioactive materials and can only be handled using thick lead shields. True or False? Please explain your choice in one sentence.
    Ans: False. Plutonium is an α emitter and α particles are relatively nonpenetrating.

85. The rate of nuclear decay does not depend on temperature. True or False?
    Ans: True

Chapter 18: Organic Chemistry I

1. Name the following compound.

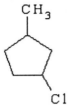

A)   cyclopentane-1-chloro-3-methyl
B)   1-methyl-4-chlorocyclopentane
C)   1-methyl-3-chlorocyclopentane

D)   1-chloro-4-methylcyclopentane
E)   1-chloro-3-methylcyclopentane

Ans: E

2. Name the following compound.

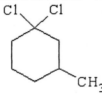

A)   1,1-dichloro-5-methylcyclohexane
B)   1-methyl-3,3-dichlorocyclohexane
C)   1,1-dichloro-3-methylcyclohexane

D)   dichlorocyclohexane-3-methyl
E)   1-methyl-5,5-dichlorocyclohexane

Ans: C

3. Name the following compound.

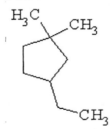

Ans: 3-ethyl-1,1-dimethylcyclopentane

4. Name the compound $CH_3CH(CH_3)CH(CH_3)CH_3$.

A)   2-isopropylpropane
B)   1-methyl-1-isopropylethane
C)   2,3-dimethylbutane

D)   1,1-dimethylbutane
E)   hexane

Ans: C

5. Name the compound $CH_3CH(CH_3)CH(CH_2CH_3)C(CH_3)_3$.
    A)   2,4,4-trimethyl-3-ethylpentane
    B)   2,2,4-trimethyl-3-ethylpentane
    C)   decane
    D)   1,2-tetramethylpentane
    E)   2,2-dimethyl-3-ethyl-4-methylpentane
    Ans: B

6. Name the following compound.

    Ans: 2,2-dimethylpropane

7. All of the following are saturated hydrocarbons except
    A) C_4H_{10}.   B) C_3H_4.   C) C_5H_{12}.   D) C_4H_6.   E) C_3H_8.
    Ans: D

8. Which of the following could be a cycloalkane?
    A) C_3H_8 B) C_5H_{12} C) C_6H_{14} D) C_6H_{12} E) C_4H_{10}
    Ans: D

9. Which of the following has the highest boiling point?
    A) C_3H_8 D) $CH_3(CH_2)_4CH_3$
    B) $CH_3(CH_2)_7CH_3$ E) $CH_3(CH_2)_5CH_3$
    C) $CH_3(CH_2)_3CH_3$
    Ans: B

10. The compound $(CH_3)_3CCH_2CH(CH_3)_2$ is
    A)   named as a pentane but is an isomer of heptane.
    B)   named as a pentane but is an isomer of octane.
    C)   named as a hexane but is an isomer of octane.
    D)   named as a pentane but is an isomer of hexane.
    E)   named as a butane but is an isomer of octane.
    Ans: B

11. Name the compound $CH_3CH(CH_2CH_3)CH(CH_2CH_3)CH(CH_3)CH(CH_3)_2$.
    A)   dodecane
    B)   2,3-diethyl-4,5-dimethylhexane
    C)   2,3-dimethyl-4,5-diethylhexane
    D)   2-ethyl-3-ethyl-4-methyl-5-methylhexane
    E)   2,3-diethy-4-isopropylpentane
    Ans: B

12. Name the compound $C(CH_3)_4$.
    A) 2-methylbutane
    B) pentane
    C) 2,2-dimethylpropane
    D) isopropylmethane
    E) isobutylmethane

    Ans: C

13. Name the compound $CH_3C(CH_3)_2C(CH_3)(CH_2CH_3)CH(CH_3)_2$.
    A) 3,4-dimethyl-3-tert-butylpentane
    B) 3-isopropyl-3,4,4-trimethylpenatane
    C) 3-ethyl-2,2,3,4-tetramethylpentane
    D) undecane
    E) 3-ethyl-2,3,4,4-tetramethylpentane

    Ans: C

14. Name the following compound.

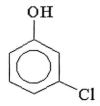

    A) 1-tert-butyl-4-ethyl-5-methylbutane
    B) undecane
    C) 1,1,1,4-tetramethyl-3-ethylpentane
    D) 3-ethyl-2,5,5-trimethylhexane
    E) 4-ethyl-2,2,5-trimethylhexane

    Ans: E

15. Name the following compound.

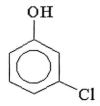

    A) 1-iodo-toluene
    B) 1-methyl-2-iodobenzene
    C) m-iodobenzene
    D) 2-iodo-3-methylbenzene
    E) m-toluene

    Ans: A

16. Name the following compound.

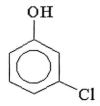

    Ans: 3-chlorophenol

17. Name the compound $(CH_3CH_2)_2CHCH=CHCH_3$.
    A)   diethylbutene
    B)   3-ethyl-4-hexene
    C)   4,4-diethyl-2-butene
    D)   4-ethyl-2-hexene
    E)   ethylhexene
    Ans: D

18. Name the compound $CH_3CH=C(CH_3)CH(CH_3)_2$.
    Ans: 3,4-dimethyl-2-pentene

19. Name the compound $(CH_3)_2CHCCCH_3$.
    Ans: 4-methyl-2-pentyne

20. How many structural isomers are possible for heptane?
    A) 9   B) 5   C) 7   D) 4   E) 6
    Ans: A

21. All of the following are structural isomers of C_6H_{14} except
    A) $CH_3(CH_2)_4CH_3$
    B) $CH_3CH_2C(CH_3)_3$
    C) $(CH_3)_2CHCH(CH_3)_2$
    D) $CH_3(CH_2)_2CH(CH_3)_2$
    E) $(CH_3)_2CHCH_2CH_3$
    Ans: E

22. How many of the structural isomers of C_6H_{14} have only primary and secondary hydrogens?
    A) 2   B) 1   C) 5   D) 4   E) 3
    Ans: A

23. How many of the structural isomers of C_6H_{14} have tertiary hydrogens?
    A) 5   B) 2   C) 1   D) 3   E) 4
    Ans: D

24. How many of the structural isomers of C_6H_{14} have only primary and tertiary hydrogens?
    A) 1   B) 2   C) 3   D) 4   E) 5
    Ans: A

25. The alkene $CH_3CH=CHCH_2CH_3$ cannot exist as *cis* and *trans* isomers.
    Ans: False

26. Enantiomers have identical chemical properties, except when they react with other chiral compounds.
    Ans: True

27. A racemic mixture is an equal mixture of enantiomers and therefore is optically active.
    Ans: False

28. Which of the following is optically active?
    A) $NH_2CH(CH_3)COOH$ D) $C(CH_3)_4$
    B) $(CH_3)_2C(NH_2)COOH$ E) NH_2CH_2COOH
    C) $CH_3CHCH(Cl)$
    Ans: A

29. Which of the following is optically active?
    A) $CH_3CHCH(Cl)$ D) NH_2CH_2COOH
    B) $(CH_3)_2C(NH_2)COOH$ E) $C(CH_3)_4$
    C) $CH_3CH_2CH(NH_2)COOH$
    Ans: C

30. Consider the compound below.

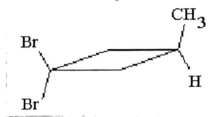

Which of the following is true?
    A) The compound has geometric isomers.
    B) The compound is not chiral.
    C) The compound exists as 3 stereoisomers.
    D) The compound is chiral and does not have geometric isomers.
    E) The compound is chiral.
    Ans: B

31. The compound 1-chloro-1-pentene
    A) can exist as *cis* and *trans* isomers.
    B) has the formula $C_5H_{11}Cl$.
    C) cannot exist as *cis* and *trans* isomers.
    D) is an alkyne.
    E) has 3 structural isomers.
    Ans: A

32. Which of the following compounds exist as geometric isomers?
            1) $CH_2=C(CH_3)(CH_2CH_3)$
            2) $CH_3CH=CH(CH_2CH_3)$
            3) $ClCH=C(CH_3)$
            4) $ClCH=C(Cl)(CH_3)$
    A) 1, 2, 3, and 4   B) 2 and 4   C) 1 and 3   D) 1, 2, and 4   E) 3 and 4
    Ans: B

33. The compound 2-chloro-1-pentene
    A)   has the formula $C_5H_{11}Cl$.
    B)   is an alkyne.
    C)   cannot exist as *cis* and *trans* isomers.
    D)   can exist as *cis* and *trans* isomers.
    E)   has 3 structural isomers.
    Ans: C

34. Name the following compound.

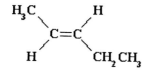

    A)   *cis*-2-pentene   D)   *trans*-1-ethyl-1-propene
    B)   *trans*-1-methyl-1-butene   E)   *trans*-2-pentene
    C)   ethylmethylethene
    Ans: E

35. Name the following compound.

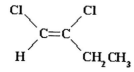

    A)   *cis*-1,2-dichloro-1-butene   D)   *trans*-1,2-dichloro-1-butene
    B)   dichlorobutene   E)   *cis*-1,2-dichloro-2-ethylethene
    C)   *cis*-2-ethyl-1,2-dichloroethene
    Ans: A

36. Which of the following have a *cis* configuration?

            1   2   3   4
    A) 2   B) 1 and 2   C) 2 and 4   D) 1, 2, and 4   E) 1 and 4
    Ans: C

37. Which of the following have a *trans* configuration?

| 1 | 2 | 3 | 4 |

A) 1, 3, and 4   B) 3   C) 1, 2, and 3   D) 1 and 3   E) 1, 2, 3, and 4
Ans: D

38. Consider a molecule of molecular formula $C_{10}H_{16}$. What total number of rings plus π bonds are present in the molecule?
A) 2   B) 1   C) 3   D) 4   E) 0
Ans: C

39. Consider a molecule of molecular formula $C_{10}H_{18}$. What total number of rings plus π bonds are present in the molecule?
A) 3   B) 1   C) 4   D) 0   E) 2
Ans: E

40. Which of the following has the highest boiling point?
A) $CH_3CH(CH_3)CH(CH_3)CH_2CH_3$ D) $CH_3CH(CH_3)(CH_2)_3CH_3$
B) C_3H_8 E) $CH_3(CH_2)_4CH_3$
C) $CH_3(CH_2)_5CH_3$
Ans: C

41. Alkanes react with
A)   concentrated sulfuric acid.
B)   the strong oxidizing agent $KMnO_4$.
C)   boiling aqueous sodium hydroxide to give alcohols.
D)   oxygen to give carbon dioxide and water.
E)   boiling nitric acid.
Ans: D

42. Alkanes do not react with boiling nitric acid because the C-C and C-H bonds are strong.
Ans: True

43. Which of the following pairs react?
A) $CH_4(g)$ and $KMnO_4(aq)$ D) $CH_4(g)$ and $HNO_3(l)$
B) $CH_4(g)$ and $NaOH(aq)$ E) $CH_4(g)$ and $H_2SO_4(l)$
C) $CH_4(g)$ and $Cl_2(g)$
Ans: C

44. How many different monochlorination products are formed when n-butane reacts with chlorine gas under the influence of UV light?
    A) 5   B) 1   C) 3   D) 2   E) 4
    Ans: D

45. How many different monochlorination products are formed when 2-methylpentane reacts with chlorine gas under the influence of UV light?
    A) 5   B) 1   C) 2   D) 3   E) 4
    Ans: A

46. The most characteristic reaction of alkenes is
    A)   addition   D)   dimerization
    B)   substitution   E)   elimination
    C)   dehydrogenation
    Ans: A

47. Arenes undergo predominantly addition reactions. True or False?
    Ans: False

48. The product of the hydrogenation of cis-2-butene is
    A) 2-butyne   B) trans-butane   C) cis-butane   D) trans-2-butane   E) butane
    Ans: E

49. The product of the reaction of 2-butene with Cl_2 is
    A)   3,3-dichlorobutane.   D)   2-chlorobutane.
    B)   2,3-dichlorobutane.   E)   2,2-dichlorobutane.
    C)   butane.
    Ans: B

50. The product of the reaction of 2-butene with HBr is
    A)   1-bromobutane.   D)   2-bromo-1-butene.
    B)   2-bromo-2-butene.   E)   2-bromobutane.
    C)   butane.
    Ans: E

51. The reaction of ethene with bromine water results in the formation of 1,2-bromoethane. This is a(n)
    A)   electrophilic reaction.   C)   substitution reaction.
    B)   nucleophilic reaction.   D)   oxidation reaction.
    Ans: A

52. When 2-methyl-1-butene reacts with HBr, what are the names of the two products formed?
    Ans: 2-bromo-2-methylbutane (major product) and 1-bromo-2-methylbutane (minor product)

53. When 2-methyl-1-butene reacts with HBr, the reaction occurs in two steps. Draw the structure of the intermediate. (Hint: Intermediate carbocations are most stable when formed at tertiary carbons.)
    Ans: $CH_3C^+(CH_3)CH_2CH_3$

54. When 2-methyl-1-butene reacts with HBr, the reaction occurs in two steps. What is the major product? (Hint: Intermediate carbocations are most stable when formed at tertiary carbons.)
    Ans: 2-bromo-2-methylbutane

55. When 2-methyl-1-butene reacts with HBr, the reaction occurs in two steps. The attacking agent in the first step is _____.
    Ans: H^+

56. Which of the following undergo nitration faster than benzene?

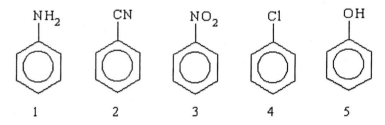

A) 4 and 5 B) 2, 3, and 5 C) 1 and 2 D) 3 and 4 E) 1, 4, and 5
Ans: E

57. Which of the following undergo nitration slower than benzene?

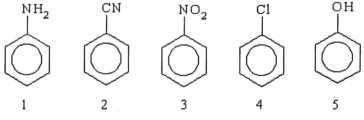

A) 3 and 4 B) 1 and 2 C) 2 and 3 D) 1, 4, and 5 E) 1 and 5
Ans: C

58. Which of the following forms ortho- and meta-products upon nitration?
    A) phenol B) benzoic acid C) nitrobenzene D) cyanobenzene (benzonitrile)
    Ans: A

59. Which of the following forms a meta-product upon nitration?
    A) benzoic acid B) aniline C) phenol D) chlorobenzene
    Ans: A

60. What major product(s) is(are) obtained from the following reaction?

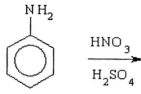

aniline

A) *ortho*-nitroaniline and *para*-nitroaniline
B) *para*-nitroaniline
C) *meta*-nitroaniline
D) *ortho*-, *meta*-, and *para*-nitroaniline
E) *ortho*-nitroaniline

Ans: A

61. What major product(s) is(are) obtained from the following reaction?

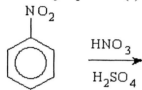

A) 1,4-dinitrobenzene
B) 1,2-, 1,3-, and 1,4-dinitrobenzene
C) 1,3-dinitrobenzene
D) 1,2-dinitrobenzene and 1,4-dinitrobenzene
E) 1,2-dinitrobenzene

Ans: C

62. How many products are predicted for the reaction below?

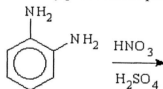

A) 4   B) no reaction occurs   C) 3   D) 1   E) 2
Ans: E

63. How many products are predicted for the reaction below?

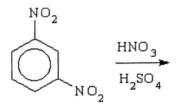

A) 1   B) 2   C) 3   D) 4   E) no reaction occurs
Ans: A

64. What major product(s) is(are) obtained from the following reaction?

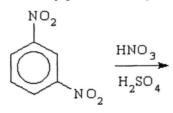

A)   1,3,5-trinitrobenzene
B)   1,2,3-trinitrobenzene
C)   1,2,3,5-tetranitrobenzene

D)   1,3,4,5-tetranitrobenzene
E)   1,3,4-trinitrobenzene

Ans: A

65. How many products are predicted *theoretically* from the following reaction?

A) 5   B) 3   C) 1   D) 4   E) 2
Ans: B

66. How many monobrominated products are predicted *theoretically* from the following
reaction?

+ Br₂ $\xrightarrow{\text{FeBr}_2}$

A) 5   B) 3   C) 4   D) 6   E) 2
Ans: E

67. What major product(s) is(are) predicted from the following reaction?

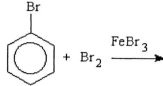

A)   *ortho*-dibromobenzene and *para*-dibromobenzene
B)   *ortho*-dibromobenzene
C)   *ortho*-, *meta*-, and *para*-dibromobenzene
D)   *meta*-dibromobenzene
E)   *para*-dibromobenzene
Ans: A

68. When benzoic acid is treated with nitric acid and concentrated sulfuric acid, the product is

A)   meta-nitrobenzoic acid.   C)   para-nitrobenzoic acid.
B)   ortho-nitrobenzoic acid.   D)   ortho- and para-nitrobenzoic acids.
Ans: A

69. What major product(s) is(are) predicted from the following reaction?

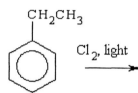

A)   1-chloro-1-phenylethane   D)   *ortho*-chloroethylbenzene
B)   *meta*-chloroethylbenzene   E)   *para*-chloroethylbenzene
C)   *ortho*- and *para*-chloroethylbenzene
Ans: A

70. What major product(s) is(are) predicted from the following reaction?

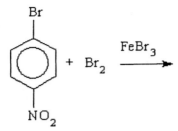

A)   The ring is deactivated and no reaction occurs.
B)   2,4-dibromonitrobenzene
C)   3,4-dibromonitrobenzene
D)   1,4-dibromobenzene
E)   1,2,3-tribromonitrobenzene
Ans: C

71. The most common reaction of benzene is addition.
Ans: False

72. Which of the following species is likely to attack benzene and result in substitution?
A) NH_3 B) $FeBr_4-$ C) OH^- D) HCl E) NO_2+
Ans: E

73. A dichlorobenzene reacts with HNO_3/H_2SO_4 and produces three mononitrated products. Identify the initial dichlorobenzene.
Ans: 1,3-dichlorobenzene

74. A dichlorobenzene reacts with HNO_3/H_2SO_4 and produces one mononitrated product. Identify the initial dichlorobenzene.
Ans: 1,4-dichlorobenzene

75. Fill in the missing reactants and products in the reaction scheme below.

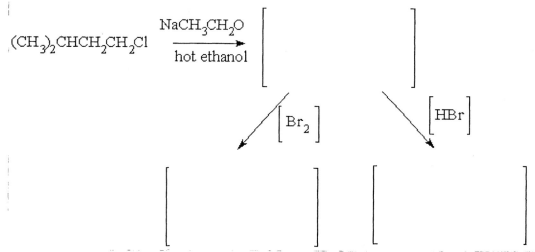

Ans:

$(CH_3)_2CHCH_2CH_2Cl$ $\xrightarrow[\text{hot ethanol}]{NaCH_3CH_2O}$ $\left[(CH_3)_2CHCH=CH_2 \right]$ + CH_3CH_2OH + Cl^-

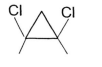 $\left[Br_2 \right]$ $\left[HBr \right]$

$\left[(CH_3)_2CHCH(Br)CH_2Br \right]$ $\left[(CH_3)_2CHCH(Br)CH_3 \right]$

76. For the compounds below, which statement is true?

$$\text{1} \qquad\qquad\qquad \text{2}$$

A)   Compound **1** is chiral.
B)   Compounds **1** and **2** are chiral.
C)   Compounds **1** and **2** are geometric isomers.
D)   Compounds **1** and **2** are identical.
Ans: C

77. For the compounds below, which statement is true?

1　　　　　　　　　**2**

A) Compound **1** is chiral.
B) Compound **2** is chiral.
Ans: B

C) Compounds **1** and **2** are chiral.
D) Compounds **1** and **2** are identical.

78. Although alkanes are quite unreactive, they do undergo substitution by a radical chain mechanism. True or False?
Ans: True

79. A mixture of nitric acid and concentrated sulfuric acid is used to convert benzene to nitrobenzene. The nitrating agent in this reaction is
A) HNO_3　B) NO_2^-　C) NO_2^+　D) NO_3^-
Ans: C

80. Consider the reaction of ethene with bromine water to produce 1,2-dibromoethane. Which of the following statements is correct regarding this reaction?
A) This reaction is a nucleophilic reaction.
B) This reaction is a substitution reaction.
C) A negatively charged intermediate is involved in this reaction.
D) A cyclic bromonium ion is an intermediate in this reaction.
Ans: D

81. Which of the following undergoes electrophilic reactions?
A) ethanol　B) butane　C) 2-propene　D) chloroethane
Ans: C

82. What is the missing compound in the reaction below?

$$\text{ethanol, 70°C}$$
$$CH_3CH_2CHBrCH_3 + \underline{\hspace{3cm}} \rightarrow CH_3CH=CHCH_3 + Br^- + CH_3CH_2OH$$

Ans: $CH_3CH_2O^-$

83. How many isomeric dichloromethylbenzenes exist?
A) four　B) five　C) six　D) seven
Ans: C

84. Consider the molecule below. Which of the following statements is correct regarding this molecule?

A) The molecule in achiral.
B) The molecule is chiral.
C) The molecule has geometric isomers.
D) The molecule exists as two stereoisomers.
Ans: A

85. Which of the compounds below is/are chiral?

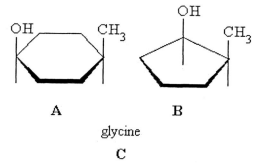

A   B

glycine

C

Ans: B

Chapter 19: Organic Chemistry II

1. Which of the following compounds is most likely to undergo nucleophilic substitution by an S_N2 reaction?
   A) $CH_3CH_2CH_2Br$
   B) $(CH_3CH_2)_3CBr$
   C) $CH_3CH_2CBr(CH_3)CH_3$
   D) $(CH_3)_3CBr$
   Ans: A

2. The compound 3-bromo-3-ethylpentane likely undergoes nucleophilic substitution by an S_N2 reaction.
   Ans: False

3. In the S_N1 mechanism for nucleophilic substitution, the carbocation produced is a Lewis acid and the nucleophile is a Lewis base.
   Ans: True

4. Where is the electrophilic site in the compound 2-bromo-2-methylbutane?
   Ans: the carbon marked with an asterisk.

   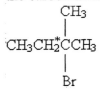

5. Consider the following reaction:
   $$CH_3(CH_2)_4CH_2CH(Br)CH_3 + NaOH \rightarrow CH_3(CH_2)_4CH_2CH(OH)CH_3$$
   (optically pure enantiomer)
   If 2-bromooctane rotates the plane of polarized light to the right while the product rotates the plane of polarized light to the left, which of the following is true?
   A)   This is an example of an elimination reaction.
   B)   The reaction occurs by an S_N2 mechanism.
   C)   The reaction occurs by an S_N1 mechanism.
   D)   This is an example of an addition reaction.
   E)   This is an example of an electrophilic substitution.
   Ans: B

6. Consider the following reaction:

   $CH_3(CH_2)_4CH_2CH(Br)CH_3 + NaOH \rightarrow CH_3(CH_2)_4CH_2CH(OH)CH_3$
   (optically pure enantiomer)

   The rate law for this reaction over a wide range of [OH$^-$] is R = k_1[RX] + k_2[RX][OH$^-$] where RX is 2-bromooctane and k_2/k_1 = 20. At very low hydroxide ion concentration,
   A)   the product of the reaction does not rotate the plane of polarized light.
   B)   the product formed is $CH_3(CH_2)_4CH=CHCH_3$ and not the alcohol.
   C)   the reaction proceeds by an S_N2 mechanism.
   D)   the hydroxide ion attacks the face of the carbocation in a concerted mechanism on the face opposite the leaving group.
   E)   the product is optically active and rotates the plane of polarized light in the opposite direction with respect to 2-bromooctane.
   Ans: A

7. If 2-bromobutane reacts with OH$^-$ by an S_N1 mechanism, the product is optically active and has the reverse chirality compared to the reactant. True or False?
   Ans: False

8. If 2-bromobutane reacts with OH$^-$ by an S_N2 mechanism, the product is optically active and has the reverse chirality compared to the reactant. True or False?
   Ans: True

9. The major product of the reaction of $(CH_3)_2CHCH(Br)CH_3$ with concentrated KOH in alcohol solvent is
   A)   2-methyl-2-butene.   D)   pentene.
   B)   3-methyl-1-butene.   E)   3-methylbuatanol.
   C)   2-hydroxy-3-methylbutane.
   Ans: A

10. All of the following function as nucleophiles in nucleophilic substitution reactions except
    A) H_2O B) $CH_3CH_2O^-$ C) SH^- D) CN^- E) PH_3
    Ans: E

11. What is the product of the following reaction?
    $C_5H_9CH_2CH_2Br + NaCN \rightarrow$
    Ans: $C_5H_9CH_2CH_2CN$

12. Which of the following is a secondary alcohol?
    A) CH_3CH_2OH D) $CH_3(CH_2)_3OH$
    B) $C_6H_5CH_2OH$ E) $CH(CH_3)_2OH$
    C) $C(CH_3)_3OH$
    Ans: E

13. Which of the following is a tertiary alcohol?
    A) $C_6H_5CH_2OH$
    B) $C(CH_3)_3OH$
    C) CH_3CH_2OH
    D) $CH(CH_3)_2OH$
    E) $CH_3(CH_2)_3OH$
    Ans: B

14. How many primary alcohols are there with the formula C_4H_9OH?
    A) 4   B) 3   C) 5   D) 1   E) 2
    Ans: E

15. How many secondary alcohols are there with the formula C_4H_9OH?
    A) 4   B) 2   C) 3   D) 1   E) 5
    Ans: D

16. Which of the following is most volatile?
    A) $(CH_3)_2C(OH)CH_3$
    B) $(CH_3)_3COH$
    C) $CH_3CH(OH)CH_2CH_3$
    D) $CH_3(CH_2)_3OH$
    E) $CH_3CH_2CH_2CH_3$
    Ans: E

17. The compound $CH_3CH_2OCH_2CH_3$ is
    A) an alcohol.   B) an ester.   C) a ketone.   D) an aldehyde.   E) an ether.
    Ans: E

18. Which of the following is most volatile?
    A) $CH_3CH(OH)CH_2CH_3$
    B) $CH_3(CH_2)_3OH$
    C) $(CH_3)_2C(OH)CH_3$
    D) $(CH_3)_3COH$
    E) $CH_3CH_2OCH_2CH_3$
    Ans: E

19. The $-OH$ group occurs in
    A) aldehydes and alcohols.
    B) alcohols only.
    C) alcohols, phenols, and carboxylic acids.
    D) phenols and ketones.
    E) ketones and carboxylic acids.
    Ans: C

20. Ethers are less volatile than alcohols of the same molar mass.
    Ans: False

21. Which of the following is the strongest acid?
    C_6H_5OH or $C_6H_5CH_2OH$
    Ans: phenol

22. There are two isomeric butanals that differ in the position of the carbonyl group.
    Ans: False

23. Which of the following compounds is the strongest acid?

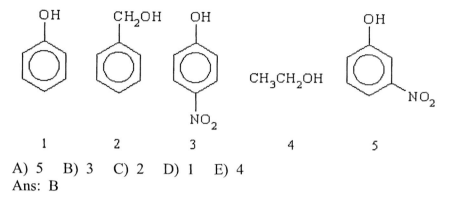

A) 5 B) 3 C) 2 D) 1 E) 4
Ans: B

24. The carbonyl group occurs in all of the following except
    A) amides.   B) ketones.   C) carboxylic acids.   D) aldehydes.   E) phenols.
    Ans: E

25. Oxidation of secondary alcohols produces
    A)   acids.
    B)   aldehydes.
    C)   ketones.
    D)   Secondary alcohols cannot be oxidized.
    E)   ethers.
    Ans: C

26. Oxidation of 2-propanol gives
    A)   acetaldehyde.   D)   dimethyl ether.
    B)   propanoic acid.   E)   acetic acid.
    C)   acetone.
    Ans: C

27. To produce butanone, which of the following should be reacted with $Na_2Cr_2O_7$(aq),
    H_2SO_4(aq)?
    A) $CH_3CH_2CH(OH)CH_3$ D) $(CH_3)_2CHCHO$
    B) $CH_3CH_2CH(COOH)CH_3$ E) $CH_3CH_2CH_2CH_2OH$
    C) $CH_3CH_2CH(CH_2OH)CH_3$
    Ans: A

28. All of the following produce a silver mirror with Tollen's reagent except
    A) HCHO.
    B) $CH_3CH_2CH(CHO)CH_3$.
    C) $CH_3CH_2COCH_3$.
    D) $CH_3(CH_2)_3CHO$.
    E) CH_3CH_2CHO.
    Ans: C

29. Which of the following produces a silver mirror with Tollen's reagent?
    A) CH_3COOH
    B) $CH_3C(O)OCH_3$
    C) CH_3COCH_3
    D) CH_3OCH_3
    E) CH_3CHO
    Ans: E

30. When aldehydes react with Tollen's reagent,
    A) a red precipitate forms.
    B) silver ions are produced.
    C) a ketone is produced.
    D) the aldehyde reduces silver ions.
    E) an alcohol is produced.
    Ans: D

31. When aldehydes react with Tollen's reagent,
    A) a red precipitate forms.
    B) a ketone is produced.
    C) silver ions are produced.
    D) an alcohol is produced.
    E) a carboxylic acid is produced.
    Ans: E

32. The product of the reaction of CH_3CH_2CHO with excess $KMnO_4$(aq) in acidic solution is
    A) CH_3CH_2COOH.
    B) $CH_3CH_2COCH_2CH_3$.
    C) CH_3COOH.
    D) $CH_3CH_2CH_3$.
    E) CH_3CH_2OH.
    Ans: A

33. What is the product of the following reaction?

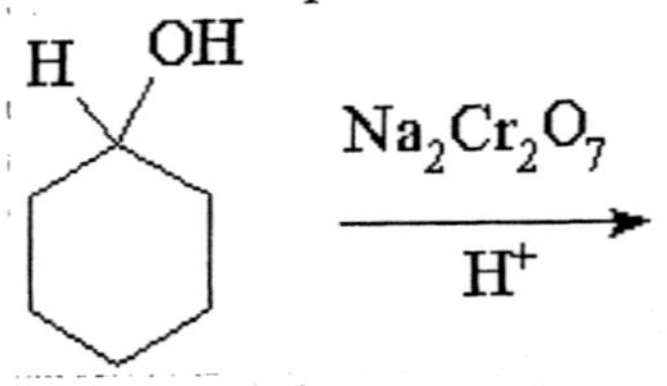

    H   OH
     \ /
    [cyclohexane ring] $\xrightarrow[H^+]{Na_2Cr_2O_7}$

    Ans: cyclohexanone

34. Which of the following can produce esters?
    A)   a primary alcohol plus $K_2Cr_2O_7$(aq), H_2SO_4
    B)   an alcohol plus an aldehyde
    C)   an aldehyde plus $KMnO_4$(aq) in acidic solution
    D)   an acid plus an alcohol
    E)   an acid plus an aldehyde
    Ans: D

35. The ester $CH_3COO(CH_2)_4CH_3$ is responsible for the odor of bananas. It can be prepared from
    A) CH_3CH_2OH and $CH_3(CH_2)_3COOH$.
    B) CH_3CHO and $CH_3(CH_2)_3CH_2OH$.
    C) CH_3COOH and $CH_3(CH_2)_3CH_2OH$.
    D) CH_3CHO and $CH_3(CH_2)_3COOH$.
    E) CH_3COOH and $CH_3(CH_2)_3CHO$.
    Ans: C

36. Name the compound $CH_3COO(CH_2)_4CH_3$.
    A)   methyl pentanoate   D)   pentyl acetate
    B)   hexanoic acid   E)   butyl acetate
    C)   2-hexanone
    Ans: D

37. The formation of an ester is
    A)   a substitution reaction.   D)   an irreversible reaction.
    B)   an oxygenation reaction.   E)   a condensation reaction.
    C)   an addition reaction.
    Ans: E

38. When an ester is formed via a condensation reaction with elimination of water, the oxygen atom in the water molecule comes from
    A)   the hydroxyl group of the acid.   D)   the carbonyl group of the acid.
    B)   the alcohol.   E)   the solvent.
    C)   the aldehyde.
    Ans: A

39. Predict the product of the reaction of acetic acid with dimethylamine.
    A) $CH_3CONH(CH_3)_2^+$ D) $CH_3CON(CH_3)_2$
    B)   No reaction occurs.   E) CH_3CONH_2
    C) $CH_3CONHCH_3$
    Ans: D

40. Predict the product of the reaction of acetic acid with trimethylamine.

A) $CH_3CON(CH_3)_3^+$

D) CH_3CONH_2

B) $CH_3CONHCH_3$

E) No reaction occurs.

C) $CH_3CON(CH_3)_2$

Ans: E

41. Which of the following would be attacked by a nucleophilic reagent?

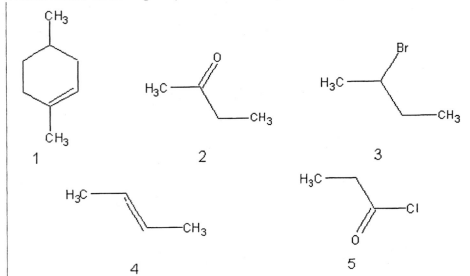

A) 2, 3, and 5   B) 1 and 4   C) 3 and 5   D) 2 and 5   E) 3 only

Ans: A

42. Which of the following would be attacked by an electrophilic reagent?

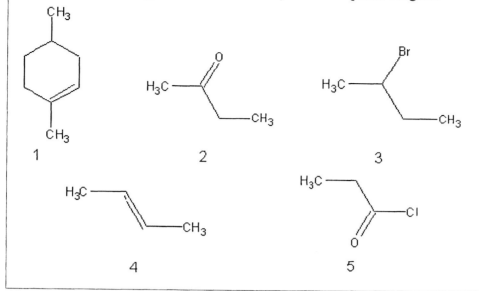

A) 2 and 5 only   D) 3 and 5
B) 1 and 4 only   E) 1 and 4 only
C) 1, 2, 4, and 5
Ans: C

43. The systematic name of $CH_3CH_2CH(CHO)CH_3$ is
    A) 2-ethylpropanal.   D) 2-pentanone.
    B) 2-pentanal.   E) 2-methylbutanal.
    C) 3-methylbutanal.
    Ans: E

44. The systematic name of $CH_3CH_2COCH_3$ is
    A) 2-butanone.   D) 2-butanal.
    B) 3-butanal.   E) 3-butanone.
    C) methypropylether.
    Ans: A

45. Which of the following reactants can be used to produce N-ethylacetamide?

CH_3COCl CO_2 $CH_3CH_2NH_2$ NH_3 CH_3CH_2COCl CH_3CONH_2
    1   2   3   4   5   6

$CH_3CH_2CH_2COCl$ $CH_3CH_2CH_2Cl$ CH_3NH_2 $HCOOH$ CH_3CH_2Cl
        7   8   9   10   11

A) 2, 9, and 11   B) 5 and 9   C) 1 and 3   D) 4, 5, and 11   E) 6 and 11
Ans: C

46. Which of the following reactants can be used to produce propanamide?

CH_3COCl CO_2 $CH_3CH_2NH_2$ NH_3 CH_3CH_2COCl CH_3CONH_2

   1   2   3   4   5   6

$CH_3CH_2CH_2COCl$ $CH_3CH_2CH_2Cl$ CH_3NH_2 $HCOOH$ CH_3CH_2Cl

     7   8   9   10   11

A) 5 and 6   B) 4 and 7   C) 2, 4, and 11   D) 4 and 5   E) 2, 4, and 8

Ans: D

47. What is the product of the following reaction?

$C_6H_5COCl + CH_3NH_2 \rightarrow$

Ans: $C_6H_5CONHCH_3$

48. What is the product of the following reaction?

$CH_3CH_2CH_2Br + NaHS \rightarrow$

Ans: $CH_3CH_2CH_2SH$

49. Which of the following monomers is used to produce Teflon?

A) $CHClCH_2$
B) CF_2CF_2
C) CH_2CH_2
D) $CH(CH_3)CH_2$
E) $CH(CN)CH_2$

Ans: B

50. Which of the following monomers is used to produce polystyrene?

A) $CH(CH_3)CH_2$
B) $CHClCH_2$
C) CH_2CH_2
D) $CH(CN)CH_2$
E) $CH(C_6H_5)CH_2$

Ans: E

51. What is the product of the following reaction?

$CH_3CH(NH_2)CH_3 + CH_3CH_2C(O)OCH_3 \rightarrow$

Ans: $CH_3CH_2C(O)NHCH(CH_3)_2$

52. Fill in the missing compound in the following reaction.

$CH_3CH_2COOH + \underline{\hspace{4cm}} \rightarrow CH_3CH_2C(O)OCH(CH_3)_2$

Ans: $(CH_3)_2CHOH$

53. Fill in the missing compound in the following reaction.

$CH_3CH_2Br + \underline{\hspace{2.5cm}} \rightarrow CH_3CH_2CN + NaI$

Ans: NaCN

54. What reactants could be used to synthesize CH_3CONH_2?

Ans: CH_3COCl and NH_3

55. The polymer that is formed from acrylonitrile is
A) Teflon.   B) Orlon.   C) Lucite.   D) polystyrene.   E) PVC.
Ans: B

56. The following polymer is called

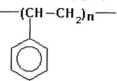

A) Teflon.   B) PVC.   C) polypropylene.   D) polyethylene.   E) polystyrene.
Ans: E

57. The following polymer is called

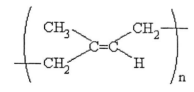

A) polyester.   B) Dacron.   C) Teflon.   D) kevlar.   E) nylon-66.
Ans: E

58. Condensation polymerization involves the use of
A)   one or two monomers, each with a nitrile group.
B)   one or two monomers with reactive groups at each end of the molecules.
C)   a monomer with a double bond.
D)   a monomer with a triple bond.
E)   one or two monomers, each with one reactive group.
Ans: B

59. All of the following polymers are produced by addition polymerization except
A) polyethylene.   B) Dacron.   C) PVC.   D) polypropylene.   E) Teflon.
Ans: B

60. The structure of the polymer rubber is

What is the formula of the monomer used to produce rubber?
A) $(CH_3)_2CCHCH_3$ D) $CH_2C(CH_3)CHCH_2$
B) CH_3CCCH_3 E) $CH_3CH(CH_3)CH_2CH_3$
C) CH_2CCCH_2
Ans: D

61. What are the formulas of the monomers used to produce the nylon polymer below?

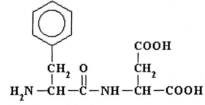

A) $NH_2CH_2CH_2NH_2$ and $(COCl)-C_6H_4-(COCl)$ (substituents are *para*)
B) $NH_2CH_2CH_2Cl$ and $(COCl)-C_6H_4-(COOH)$
C) $HOOC-CH_2CH_2-COOH$ and $NH_2-C_6H_4-NH_2$ (substituents are *para*)
D) $ClCH_2CH_2Cl$ and $(COOH)-C_6H_4-(COOH)$ (substituents are *para*)
E) $ClCH_2CH_2Cl$ and $NH_2-C_6H_4-NH_2$ (substituents are *para*)
Ans: A

62. Which of the following is cysteine?
A) $CH_2(NH_2)COOH$ D) $HSCH_2CH(NH_2)COOH$
B) $CH_3CH(NH_2)COOH$ E) $(CH_3)_2CHCH(NH_2)COOH$
C) $HOCH_2CH(NH_2)COOH$
Ans: D

63. Which of the following is a peptide bond?
A) $-C(O)O-$ B) $-CONH-$ C) $-C-NH-$ D) $-C=N-$ E) $-C(O)-$
Ans: B

64. Which of the following refers to the secondary structure of a protein?
A) denaturation   D) disulfide linkages
B) α helix   E) base pairs
C) amino acid sequence
Ans: B

65. How many different tripeptides can be formed from 3 different amino acids?
A) 3   B) 4   C) 6   D) 2   E) 12
Ans: C

66. The primary structure of aspartame is shown below.

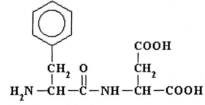

This dipeptide is
A) Phe-Asp.   B) Trp-Glu.   C) Phe-Glu.   D) Tyr-Glu.   E) Tyr-Asn.
Ans: A

67. The α helix results from the pairing of specific bases by hydrogen bonding. Which pairs form hydrogen bonds?
    A) AT and GC
    B) AC and GC
    C) GT
    D) AT, GC, AC, and GT
    E) AC
    Ans: A

68. If the base sequence along a portion of one strand of a double helix is CTACACG, the corresponding sequence on the other strand is
    A) GATGTGC.
    B) CUACACG.
    C) TCGTGTC.
    D) CTUCUCG.
    E) CTACACG.
    Ans: A

69. How many hydrogen bonds are formed between T and A?
    A) two   B) one   C) three   D) four
    Ans: A

70. The primary structure of a peptide is shown below.

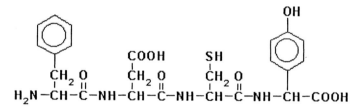

    This peptide is
    A) Tyr-Asn-Cys-Phe.
    B) Tyr-Glu-Cys-Phe.
    C) Phe-Asp-Cys-Tyr.
    D) Phe-Glu-Met-Tyr.
    E) Tyr-Asp-Met-Phe.
    Ans: C

71. Which of the following polymers are condensation polymers?

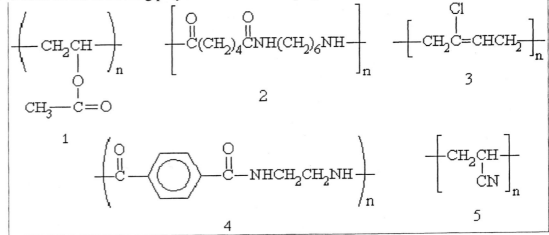

A) 3 only.   B) 2 and 4.   C) 2 and 3.   D) 4 and 5.   E) 1, 3, and 5.
Ans: B

72. Which of the following polymers are addition polymers?

A) 1, 3, and 5.   B) 2 and 3.   C) 2 and 4.   D) 4 and 5.   E) 3 only.
Ans: A

73. What reactants could be used to synthesize $CH_3CH_2CH_2C(O)OCH_3$?
Ans: CH_3OH and $CH_3CH_2CH_2COOH$

74. Consider the following reactions.
$$CH_3CH_2CH_3 + Br_2/light \rightarrow A$$
$$A + NaCN \rightarrow B$$
Identify A and B.
Ans: A is 2-bromopropane and B is 2-cyanopropane.

75. Fill in the missing reactants and products.

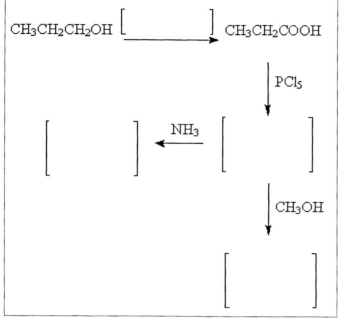

Ans:

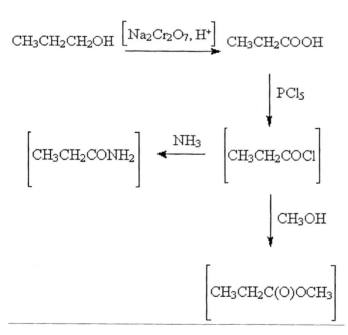

76. Esters can be produced by a condensation reaction between acid halides and alcohols. If the acid halide CH_3COCl reacts with ethanol, which of the following is eliminated?
A) HCl   B) H_2O C) OH^- D) H_3O^+
Ans: A

77. When amines condense with carboxylic acids what molecule is eliminated?
Ans: H_2O

78. In the reaction of CH_3CH_2COOH with CH_3NH_2 what molecule is eliminated?
Ans: H_2O

79. Amines react with both carboxylic acids and acid chlorides to produce amides. True or False?
Ans: True

80. What reactants would be used to synthesize the following compound?

Choose from the following.

Ans: 2 and 3

81. What reactants would be used to synthesize the following compound?

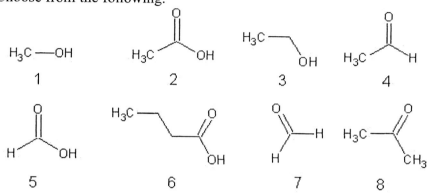

Choose from the following.

H_3C —OH

1

H_3C $\overset{O}{\overset{\|}{C}}$ OH

2

H_3C —\\—OH

3

H_3C $\overset{O}{\overset{\|}{C}}$ H

4

H $\overset{O}{\overset{\|}{C}}$ OH

5

H_3C —\\—\overset{O}{\overset{\|}{C}} OH

6

$\overset{O}{\overset{\|}{C}}$ —H H

7

H_3C $\overset{O}{\overset{\|}{C}}$ CH_3

8

Ans: 3 and 5

82. What reactants would be used to synthesize the following compound?

H_3C — $\overset{O}{\overset{\|}{C}}$ — NH — \\ — CH_3

Ans: acetyl chloride and ethylamine

83. Acetic acetyl chloride reacts with methylamine to produce _____?
Ans: N-methylacetamide

84. What reactants would be used to synthesize acetamide?
Ans: acetyl chloride and ammonia

85. Kevlar is produced from HOOC-C_6H_4-COOH and H_2N-C_6H_4-NH_2. Write the formula for Kevlar.
Ans: $(CO$-C_6H_4-$CONH$-C_6H_4-$NH)_n$